New Horizons in Standardized Work

Techniques for Manufacturing and Business Process Improvement

New Horizons in Standardized Work

Techniques for Manufacturing and Business Process Improvement

Timothy D. Martin & Jeffrey T. Bell

CRC Press
Taylor & Francis Group
Boca Raton London New York

CRC Press is an imprint of the
Taylor & Francis Group, an **informa** business

A PRODUCTIVITY PRESS BOOK

Productivity Press
Taylor & Francis Group
270 Madison Avenue
New York, NY 10016

Library of Congress Cataloging-in-Publication Data

Martin, Timothy D.
   New horizons in standardized work : techniques for manufacturing and business process
Improvement / Timothy D. Martin and Jeffrey T. Bell.
       p. cm.
   Includes bibliographical references and index.
   ISBN 978-1-4398-4080-1
   1. Workflow. 2. Standardization. 3. Production standards. I. Bell, Jeffrey T. II. Title.

HD62.17.M37 2011
658.5'3--dc22
                                                                                    2010028757

Visit the Taylor & Francis Web site at
http://www.taylorandfrancis.com

and the Productivity Press Web site at
http://www.productivitypress.com

Contents

Foreword

The statement, "Without Standards, there can be no improvements," is one that I became well aware of during my tenure at Toyota. During my time as a manager in the assembly plant at the Toyota Motor Manufacturing Kentucky facility, I became very reliant on standard work to not only effectively produce cars, but to continually improve the process that we used to produce those cars.

A common misconception about standardized work is that it is only for the person doing the standardized work. The person that does the standardized work every day knows standardized work. Standardized work is for management. Standardized work allows the management of an organization to verify that the process is being done correctly and provides opportunities for improving the process. Standardized work is vital to the continuous improvement efforts in a Lean Enterprise System.

In this book, Tim and Jeff walk you through a methodology that they have developed to effectively create and develop standardized work. You will find as you read this book that to effectively create standardized work, you will have to go to the floor and "sweat the details" of your process. If (and I do mean If) you are dedicated to creating good standardized work, you will certainly gain detailed knowledge of your processes, which is a knowledge that most of us have not had in the past.

Process knowledge in many organizations is held by a select number of people who are involved in the process, and each of them likely has a little different bit of knowledge. In other words, valuable process knowledge is only provided on a "need to know" basis. When process knowledge is held in this manner, improving the process and removing waste is very difficult, because knowledge is hard to gain and there is no true standard to improve.

By going through the process of creating standardized work that Tim and Jeff have outlined in this book, you can go from having process knowledge that is only held by the two or three people, to being able to post the correct standard work for doing the work every day. When a facility has this type of standard work, they are equipped with a necessary variable to continually improve their Lean Enterprise System.

There can be no doubt that standardized work is a base for a solid Lean Enterprise System. To continually improve your operating systems, there has to be a base to begin, and standardized work is that base.

I sincerely wish you the best on your Lean journey.

Best wishes,

Rick Harris
President, Harris Lean Systems

Preface

Why write a book just on standardized work? My main reason for writing this book is to share what my coauthor and I have learned and observed in an effort to help others. You may not agree with the approach taken, but please bear with us, as this is how we have come to understand standardized work and thus must convey the points we feel are significant in a way that we feel most comfortable with. There are also several appendices where we have tried to provide some additional information in order to offer more support for certain issues that will arise as you expand your thinking about standardized work. We hope that you find them helpful as well.

One of the main points that we hope to make with this work is that it is imperative that you have a solid understanding of standardized work from a cyclic perspective before you try to adapt it to other applications. The reason for this is that understanding the fundamentals, which are derived from the application of standardized work to simple cyclic work situations, is the key to adapting them to other, sometimes very different, situations. Once this firm understanding is established, it can be used as the anchor point or foundational basis for applying standardized work to other work situations that may never have seemed possible before. Examples include health care, construction, business processes, and even food services. We have used a lot of examples that are based on manufacturing, but it is only because this application is so prevalent and is the basis of the origin of standardized work as we know it today. In other words, if you have a solid understanding of standardized work principles and the thinking behind their development, you can successfully extend and adapt them to totally different work situations and take standardized work into other areas: It is not just for manufacturing!

My journey to try and understand standardized work began many years ago. The company I was working for at the time was in the process of trying to implement "Lean" manufacturing. It was quite a large corporation, and throughout the company they utilized outside consultants to try and teach us about this philosophy and how to adapt it to our businesses. I was working in manufacturing engineering and had been struggling with how we could be more competitive with all the foreign companies causing great pressure for us to compete or get out of the market. I was very interested in Dr. Goldratt's Theory of Constraints (TOC) and spent a considerable amount of time and effort, both at work and at home, learning and reading all that I could find on the subject and applying the principles to our manufacturing operations.

We were having good results (at least they were very promising), and we were expanding our implementation in the manufacturing area where I was assigned. About this time was when I was first exposed to the concepts of Lean. Because I was somewhat of a proactive maverick (many of my colleagues over the years might use much more colorful descriptions), I was transferred to a department where manufacturing engineering was going to adapt Lean into the process where we designed production systems for new projects around the world. The idea was to try and develop a Lean system before the equipment was ever purchased so that we could place Lean systems into production from the start. At this point, I should mention that the company was linked to the automotive industry, and the products tended to be complex, high volume, and very unique, making this activity very important, since the capital equipment costs for our industry were substantial.

I was eventually assigned the responsibility to help our teams integrate Lean into the production equipment. After several years trying to champion this idea with limited success, I was fortunate enough to gain learning opportunities from some very knowledgeable sources. One of my earliest and best teachers was Rick Harris. Rick had a very unique background: Not only had he worked for Toyota for many years, he had actually worked in our industry prior to working for Toyota and therefore could see both sides of the problems we were facing. I learned a great deal while working with Rick and attribute my passion for the subject to his great coaching style and infinite patience. Rick showed me that although we were not Toyota, and never would be, we could still learn *why* they did the things the way they did in the Toyota Production System (TPS) and *adapt* them to our business. This is in contrast to the approach that many companies seemed to be taking at the time, which was: Imitate Toyota and hope to receive the same results. With Rick's help, I was able to begin looking at things from a vastly different perspective, and this is when my journey truly began. I worked with and learned from Rick for several years.

As time passed, and I learned more through various projects, I eventually had the opportunity to work on a product for Toyota. This was probably the defining point in my journey to learn Lean. At this point I had originally intended to express my appreciation to several of my friends at Toyota because I had learned so much from them. However, upon reflection, I decided it was best just to express my appreciation to Toyota for their willingness to share freely with the world what they have learned and developed over the years in their relentless pursuit to eliminate waste.

Finally, back to why we are writing this book. We cannot "tell" you how to learn about standardized work. We can only try to share with you what we have learned and hope that as you try to apply the philosophy and principles on the shop floor or in the office or other place of business (wherever the *real* work happens), you will learn by doing. However, as we have adapted TPS thinking to other business processes, it has opened new doors as to how to apply the concepts that TPS is based upon. We do not claim to have all the answers, and we continue to learn with every opportunity. But we have heard so many people say that these concepts cannot work in their industry or with their processes—especially for standardized work, one of the foundations of the TPS house—that we just had to try to find a way to share what we have learned. Hopefully you will find this book helpful in your journey.

Timothy D. Martin

I have been very fortunate as well with the people that I have met in my journey learning about and implementing standardized work. I have had a number of opportunities to work with Rick Harris, both as an industrial engineer and as an industrial engineering supervisor. Rick was consulting with our company in various forms, both in current production and what we call "pre-production." Having trained and worked in industry as a "classical" industrial engineer for 25+ years, I began to really "see" the problems with Rick's help. In the early stages, we looked at the high level through value-stream mapping. This gave me a better insight into understanding human, machine, and material interactions from a "why" perspective. He further enhanced my ability to "see" at the detailed level or standardized work level of human, machine, and material. Rick had really helped us with the effective use of workplace mock-ups for both the production and pre-production settings to help troubleshoot problems and train individuals about standardized work. Many thanks to Rick for his patience and insight into tackling large company structures.

I would also like to take this opportunity to thank my coauthor, Tim Martin. He has been a great colleague, teacher, and friend in many different ways. He helped me bring out the preconceived notions and bias that I had about standardized work. Without Tim, this book would not have even been a dream.

Tim and I have spent several hundred hours over the years between ourselves and our colleagues in the discussion, agreement, and debate about standardized work. I, like Tim, truly hope that this book will help you to learn how to see the real issues in the workplace and how standardized work can help you be successful in your endeavors. Oh, and hopefully you will have some fun at the same time.

Jeffrey T. Bell

Acknowledgments

We would also like to say a very special thank you to several friends and family members whose support, suggestions, and feedback were essential in the completion of this project. Without them this book would never have been a reality.

Carla Martin
Amber Jordan
Mike Reprogle
Harold Redlin
Trevor Harris
George Bell
Marjorie Bell
Brian Summerton

Professor Henry Kraebber
Rick Harris
Chris Harris
Professor James Barany
Jimmy Martin
Margene Martin
Kimberly Martin

Chapter 1

What Is Standardized Work?

Before jumping into standardized work, it is probably a good idea to mention the Toyota Production System (TPS) and Lean manufacturing (sometimes just referred to as Lean). These topics can fill entire volumes, so there is no way we can hope to successfully cover all the principles and the philosophies behind them in this book. Therefore, the assumption is made that you are already learning about them and have at least a basic understanding of the concepts. If you have not started learning about Lean, we highly recommend that you do this before venturing further into this book.

The first thing that we would like to discuss is that standardized work is not just a format for documentation of work. It is a basic TPS philosophy that is intertwined with that of kaizen, which creates a continuous improvement environment by providing a constant "pull" to make things better, though in very small increments. With this relationship, it becomes quite obvious that standardized work is not permanent, even if the current process seems optimum. There must be a constant drive for improvement, but to minimize risks, the improvements should be in very small steps so that they can be quickly evaluated so that the gains can be established as the next norm or the changes abandoned with the least disruption. In this book, we will refer to this as *kaizen* rather than continuous improvement because we believe that there is much more to the concept than periodic improvements (large or small). The concept is that the normal condition is not a permanent situation but only a stable environment during the constant design, evaluation, and implementation of small increments of positive change for the situation that moves it to better support the company's goals. We came to understand this as the kaizen attitude—the attitude of constantly looking for ways to improve. It works best in a culture where everyone is working towards a common goal by moving the company in the same basic direction. When this occurs, it can help to establish an environment that assists everyone in the company in making decisions that help support the overall company direction. In this context, we will consider the kaizen attitude to be a basic premise that helps complete the philosophy of standardized work and continue on to developing the definition.

Standardized work can be defined as the currently best-known method for accomplishing the work. This assumes that it is the safest and most efficient method to do the work that meets the required level of quality. It is very important to understand that these assumptions are critical to learning about standardized work. Although they may seem to be self-explanatory, they also have a great effect upon the way that we think about and develop standardized work. For example, consider the assumption about the required level of quality. It is always sound thinking to strive for the highest quality products, but there is a point where this crosses the customer's perceived *value* of the product when it drives the cost too high to be profitable at the market price and we try to raise the selling price to offset this. Examples of this might include greater accuracy or functionality than is required by the customer, much better surface finish, or more expensive materials than are needed to meet the minimum customer requirements.

There are always trade-offs in the decisions that we make. If the cost to achieve a level of quality is too high, there can be no profit, since the market determines the selling price (market price − cost = profit). However, if the customer perceives our quality to be lower than that of our competitors, this can also negatively impact our business. Therefore the level of quality needs to be as high as we can get it and still remain competitive. We must constantly strive to increase our quality or we risk losing market share to our competitors.

Over the years we have heard many versions of an old joke about two hikers and a bear. Basically, a pair of hikers is walking along a trail in the woods when they see a large grizzly bear running toward them several hundred yards ahead. The first hiker starts running in the opposite direction, then notices that his companion is not running along with him. He turns to look back and sees that the second hiker has stripped his hiking boots off and is quickly donning a pair of running shoes from his backpack, which is now on the ground. He yells to his companion, "You don't think you can outrun a grizzly bear do you?" His companion replies, "I don't have to outrun the bear, I only have to outrun you!"

There is a bit of common sense to the second hiker's philosophy. If we compare this story with the concept of perceived quality, we see that we do not have to have the highest quality levels physically achievable. We only need to be higher than our competitors. The point to realize is that quality is not just a number that can be measured but rather a perception. This perception also extends to the concept of customer value. TPS and Lean manufacturing teach us about the various forms of waste. One of these forms of waste is sometimes referred to as the waste of processing. If the customers do not perceive the value in the extra processing, then how can we expect them to pay for it? This is why we cannot normally pass on our costs to our customers. If we recall the concept of value, then one way of describing value is: what the customer is willing to pay for. It follows then that anything that is left is what *we* pay for as a supplier of the good or service and really just represents waste that adds unnecessary cost to the product or service, which reduces our profit opportunity.

A Foundation Based on Stability

The "Toyota House" has been shown in many different forms over the years. It is normally expressed as a house, with a foundation and two pillars supporting a roof or ultimate goal. The main point that we want to focus on is that the foundation is based on stability (Dennis 2002, 18). Before the house can stand for the long run, the foundation must be firm and stable. This concept of creating a stable foundation or base is a central theme for the concepts of standardization and continuous improvement.

Now we return to the original definition: the currently best known method for accomplishing the work. The word *current* implies that this is how the situation is now but may change in the future. This is quite true, since one of the main principles of TPS is kaizen, or the attitude of constantly looking for waste and eliminating it. But we have learned not to wait for large (home runs) improvements or even multiple simultaneous improvements, but rather to look for small ways to improve constantly, *day-in and day-out* (base hits). It is important to note that often there are many seemingly conflicting principles in TPS. One that appears at this point is the fact that we standardize as a way of stabilizing and thus reducing unevenness or variation. However, the concept of kaizen teaches us that we must constantly strive to improve, which means change, and we are no longer stable, since the change will result in variation. The main issue in such situations is to look at things from a different perspective. The end *result* is not to stabilize (and achieve the best situation—thus declaring victory and marching on to

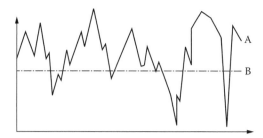

Figure 1.1   Production output under significantly different conditions.

our next problem) but to constantly look for more small improvements. We cannot wait until we are in a crisis before starting to seek improvements. We should make the kaizen attitude a part of our overall culture. But, if we first look to stabilize the situation, we can see that the more stable the situation, the smaller the incremental changes that can be implemented and verified to see whether they improve the situation or not. Consider Figure 1.1.

Suppose that the two lines, A and B, on the graph represent the output of a production department under two significantly different operating conditions. One could easily see that the average of A is much higher than B. However, in the case of A, it is much more difficult, if not impossible, to tell if a certain action actually improves the output unless the change was extreme. The reason for this is obvious: variation from cycle to cycle (in this case day-to-day). With so much variation, only time could tell if a small change made the situation better or worse. What we are trying to say is that there is obvious variation under the conditions associated with line A, and only large changes might get recognized as having an impact in the short term without looking at the average over a long period of time. This could easily result in considerable wasted effort and resources.

However, in the case of B, it is much easier to see if a small change has an impact on the output, since it is stable. Consider the amount of resources and attention that are required to manage a system with output like A. Effort and resources are required for monitoring and responding to many problems and conditions. The output of B, although lower than the average of A, is then much more receptive to applying the concept of kaizen (small, constant, incremental improvements). There is a temptation to try to go into a situation like A and just *radically* change things. Unfortunately, this is sometimes what it takes to get things stable in some severe cases. But it is often very expensive and time consuming—and what if the results are not better, or even worse? It is not a very pleasant feeling to have to explain to management that all the resources and money spent to make an improvement did not pan out! So, the old adage of changing for the sake of change without understanding the impact of the change holds true.

One of the many things we like about the TPS approach is that if changes are kept small, it is a simple matter to determine what to expect as far as results. Since the changes are much simpler, often you can actually try them out on a temporary basis before a permanent change by simulating the change right at the place where the work happens. If the results are not as expected, you have not expended a lot of money and resources. Even if the change cannot be simulated without actually implementing it, usually it can be undone quickly when it is kept small. However, do not forget that, in the previous example, the system output of B is still much less than the output of A. The kaizen attitude, if present, will provide the drive for improvement. If the changes are simple but constant, we can still make rapid improvements even when doing so in small incremental steps.

The Best Method at the Moment

Returning again to the definition, the currently best known method for accomplishing the work, we next consider the best known method. It should go without saying that the best *known* method implies that there may be a better method out there somewhere. However, we feel that it is important to mention it because it is one of the underlying principles that provide the momentum behind the kaizen attitude. We have observed many applications where the system was considered by the team to be "Lean," and after implementation there was assumed to be nothing more to be done. After all, there are only so many resources to go around, and the "Lean" experts can only be in so many places at one time, right? If consensus is that nothing more needs to be done at the moment, the kaizen attitude is not present, and the opportunity to continue to improve is being lost. This is probably one of the most frustrating situations that we have observed. It often happens when "Lean" activities are looked at as tools or events rather than as a culture.

Take for example the so-called kaizen blitz approach. There is nothing wrong with such an approach or event. Quite to the contrary, these can be defining events when trying to make a major culture change in an organization. And they can be rapid and extremely successful. But it is essential to get everyone on board and engaged in learning and improving. Granted, to be competitive, we cannot afford unlimited resources, so it is important to make the most out of the resources available. The most important resource that is often overlooked is the people who actually do the work. While observing Toyota, we noticed that the associate or team member plays a significant role in the development and improvement of standardized work. After all, who better understands the issues involved than the people who are doing it day after day? This provides them with an "army" of resources, all focused on the helping the company meet its goals! It can make a big impact on your competitiveness if your entire company is constantly looking for ways to improve.

A Basis for Improvement

Finally, the last part of the definition refers to the method for accomplishing the work. The word *method* implies that we want to *repeat* the manner or way that the work is done each time. It follows then that there is a *sequence* to the steps that constitute the method. As we will soon see, the concepts of *method* and *work sequence* will play important roles in this approach to standardized work.

So where does this lead? The obvious conclusion is that standardized work defines the work sequence to be repeated in order to try and achieve repeatable levels of safety, quality, and output. But it also shows us that it has two main usages:

1. It provides a standard method for accomplishing the work at the moment.
2. It serves as a baseline for the kaizen activities.

This brings us back to the appearance of conflicts mentioned earlier. We standardize in order to constantly improve, which means continuous change! Actually, this is somewhat of an overstatement. The first step is to stabilize the situation, as we discussed in the earlier example of the graph of output B. Standardization helps us to do this. After the situation is stable, we can now start on the road to constantly improving, but in small steps, avoiding the risk and uncertainty that is involved with large and complex changes. This will also allow us to maintain the integrity of the previous kaizen improvements.

How Do We Get Standardized Work?

First, there are some prerequisite conditions (or preconditions) that must be present in order to achieve and support standardized work. These are:

1. *A human must be capable of doing the work.* In other words, it must be work that a person can do safely and ergonomically within the time required and at the desired level of quality.
2. *There must be a repeatable sequence to the work.* The work that is required must be such that the worker can do it the same way each and every time it is expected.
3. *The equipment, tools, and workplace must be highly reliable.* If any of these are not reliable or cause problems, either in time variations per cycle or frequent downtime situations, the work will be impacted greatly.
4. *The materials used must be of high quality.* If the material and components that are used are not of good quality, there may be frequent instances of defects and other quality problems that will be a constant source of issues that cause variation that will disrupt the worker.

These preconditions are necessary in order to make sure that variability is kept to a minimum. For example, if a worker is constantly dealing with equipment downtime or bad materials, how can he or she be expected to accomplish the work in a consistent manner? Consider the variation in a task when a worker has to constantly remove excess plastic flash from an over-molded component: How can this be standardized if only a portion of the parts have excess flash or if the time required to remove the flash varies greatly between parts or between workers? Although a standard method of ensuring that the component does not move on to the next process with excessive flash might possibly be developed, resources would probably be better utilized in improving the process causing the variation in flash— whether internal or external to the company. Therefore, there is a need for the preconditions to exist first in order to establish the stability needed to support standardized work and kaizen.

At this point, we want to stop for a moment and talk briefly about the term *worker*. It is important to understand that everyone has work that they do, and therefore we use this term to refer to the person performing the work. This could be a CEO of a company, a policeman, a person assembling parts in a factory, an engineer creating new product designs, or a person preparing a meal in a restaurant. The term is not meant to imply any disrespect to people, as we have come to understand that people are the ones performing the real work—those tasks that are actually adding the value in advancement of the product or service toward completion. Machines, robots, and other complicated mechanisms are still just tools used by the worker to accomplish the work. We use the term *worker* because it is the most descriptive, and therefore the chance of miscommunication is eliminated. There is an important relationship that develops when the kaizen attitude is present. This is the symbiotic relationship between the worker and the company. When both are working toward the same goals, the results can be greatly increased, but if there are problems such as mistrust, fear, or other negative perceptions, this can prevent the two from working toward the same goals. We have also observed this unfortunate situation, and it is often one of the underlying causes of variability in some companies and should be a main focus for gaining stability. The kaizen attitude and a few of the issues involved are discussed further in Appendix E.

The Required Components for Standardized Work

Once the necessary preconditions for standardized work exist, we next need to discuss the basic components that must always be present in one form or another. These have been described by Ohno (1988, 22) as:

1. Cycle (takt) time
2. Work sequence
3. Standard inventory (in-process stock)

Please note that as we describe these required components, we will mostly use examples that relate to manufacturing. This is critical because the concepts and principles used to develop the philosophy of standardized work have their foundations in the manufacturing industry. However, it is important to understand that these required components can also be applied in other industries or business applications if we grasp the basic concepts firmly. It is this last premise that the later sections of this book are based upon in order to make the progression from manufacturing to completely different applications and industries. So please have an open mind as you consider how these required components apply to your business.

Cycle time in this context has been more recently described as takt time (Dennis 2002, 51) and is a concept that is used to describe the cadence or pace at which a product (or service) needs to be produced in order to satisfy customer demand in the time we have *allotted* to do so. The basic formula is very simple:

Takt time (TT) = Allotted time ÷ customer demand

This can be for any time period as long as the time and demand used are aligned (e.g., allotted time per day divided by customer demand per day). In the manufacturing industry, it is most commonly expressed in units of "time per part." The customer needs a certain number of parts, and we have set aside a certain amount of time to produce them. This could also be considered as somewhat of a schedule, except that it is used to act as a cadence to compare the cumulative output to the passage of time. This is a somewhat stranger definition than the conventional one, but it is important to understand that takt is a much broader concept than is apparent on first glance. It is this broader scope that will be necessary to apply the concept of takt to areas where it did not seem applicable before. We will look at this concept more deeply as we discuss the issues leading to noncyclic applications in later sections.

Work sequence refers to the sequence of steps that must occur each time the work is performed. This is a pretty straightforward statement, but the emphasis is on the sequence rather than a checklist of events that must occur without regard to the order. Whenever the work is to happen, it is important that it happen in the same sequence each time. We see now that variation can have a definite impact on many aspects of the business, especially in nonmanufacturing environments such as health care, food services, etc. However, when we begin to consider the method used to get the work done, it is not just the steps but the *manner* (often very specific motions or actions) in which each step is performed that is important. Consider that visual control, another philosophy of TPS, can help ensure that an abnormal condition (such as not following the prescribed method) is very apparent. Therefore, if there is a definite manner in which the work is to be performed, it is easier to notice abnormalities in the sequence. When we are trying to change to a culture that fosters the kaizen attitude, it is sometimes necessary to institute policies to help ensure a complete transition to the new environment. Visual control is an important concept that can assist in this transition. For more discussion on the topic of auditing standardized work, see Appendix A.

Standard inventory, otherwise known as standard in-process stock (Dennis 2002, 52), refers to any stock or material that must be present in order for the work sequence to occur as *intended*. For example, if the sequence defines that the first step is to place a part in a tooling fixture, it means that there should not be a part already in the fixture when the worker gets to that step. If there is, the sequence is disrupted and cannot be repeated until each step has the proper in-process stock (whether material is required to be present or not present at a particular location). This can be somewhat compared to the priming of an old-style hand-operated well pump. It was necessary to "prime" the pump with water in order for the pump to work as required to supply water. So in this context, it is essential that the standard in-process stock be present (or not present) according to the intended condition. This also implies that once a cycle is started, it should continue until it is complete to provide the worker with a consistent starting point after scheduled stops such as breaks, lunch periods, end of shift, etc.

One of the main reasons for in-process stock is in order to take advantage of some of the machine's automation capability, which can reduce or eliminate some of the worker's time being wasted waiting on a machine cycle to complete. For example, if a process takes 10 seconds of machine time to complete, if the job is designed so that a completed part from the previous cycle is present and ready to be exchanged with a completed part from the previous process that the worker brought with him (the completed part unloaded and the part brought with the worker loaded in its place), all that is necessary is this exchange and the activation of the machine cycle. The worker can then leave this workstation and take the completed part that was unloaded to the next process in the work sequence. If the job was designed without a completed part present, the worker would have to wait for 10 seconds before the completed part could be unloaded and moved to the next step (because the process is designed to be empty when the worker approaches on the next cycle). Common occurrences of this type of wait are when the worker must use both hands to activate the process and keep the buttons pushed in the entire time the machine is cycling for safety purposes. If the machine has ample safety features to allow the machine to be activated and free the worker of the wait, the worker can continue, and a completed part will be present when the worker returns on the next cycle if the job is designed properly. Therefore, we see that standard in-process stock (SIP) is very important and that there is a relationship between the operation of the machine and the standardized work.

One last point to mention, we notice that work sequence is both a precondition as well as a required component of standardized work. This is because it is the premise on which the principle is founded. Takt time, for setting the required pace, and standard in-process stock, to set the required materials for repetition, complete the components required for standardized work to be possible.

Types of Standardized Work

Now that we are familiar with the three required components that must always be present in some form, it is time to discuss the basic categories into which standardized work is usually broken out. For most applications where standardized work is said to be applied, they can be roughly separated into two sections: those where each product or service is basically the same and those that are different each time they are produced. The first section can be further broken down into those that pretty much require the same process steps, work, and process times (even if there are some slight differences such as color, programming, etc.), and those that are essentially the same but with some differences between models (added or deleted process steps, process-time differences between models, etc.). This results in three common categories into which most applications of standardized work are normally considered to fall.

Common categories of standardized work:

Category A: Standardized work that has the same work sequence, work content, and cycle time for all products in the same system. In other words, all the products are basically the same or take the same amount of time and same process steps, with no real differences other than things like different color components, different programming, etc., so that the only real issue is that the different parts only need to be kept separate for delivery to the respective customers. This is probably the most common category of standardized work. Since essentially the products are all the same and take the same amount of resources to build, the total number of products can be added together and considered as a single product when determining a takt time as well as for purposes of determining system capacity, even though there may be some changeover time between the different products.

Category B: Standardized work that may have almost the same basic work sequence, but the work content and therefore the cycle times are not the same between different products produced by the same system. There may be differences in the time per process step, added or skipped steps, or other variations that differentiates one product from another. A good example of this is a car on an assembly line where one unit has a sun roof and the next unit does not, even though nearly everything else may be the same. Therefore, the amount of time and resources taken to produce them in the system is not the same between products, preventing the totals from simply being added together for a common takt time calculation. This can add complexity in trying to establish time for the pace to be used as a standard. This takt time calculation problem is due to the fact that the total number of products cannot be simply added together and divided by the allotted time like the Category A standardized work application, since the products take different amounts of times to complete. Even if a single time is developed for use while these parts are being produced, it will not be very helpful unless a lot of resources are expended in comparing the output to the passage of time, since the relationship is not directly proportional as it is with Category A applications. The greater the difference in the work content between the products, the more sensitive the aggregate or weighted time is to a change in the product mix ratio. This tends to result in multiple "individual" takt times and the products being run in batches, since a single takt time that was an average would be so difficult to compare with the passage of time. Therefore, this category often causes a lot of problems with determining resources required and maximum capacities, since these are also greatly affected by the mix.

Category C: Standardized work that does not appear to have a repeatable work sequence, at least within a single work period in the same system. This category of standardized work has often been the most difficult one to successfully implement. This is because the concept of takt time does not seem appropriate because the required work is not predictable in occurrence and frequency like Category A standardized work. In most instances, this type of standardized work is not considered stable or repeatable.

For our purposes, these categories are not sufficient to properly describe the techniques and principles that will be discussed. In this book, we will discuss two main types of standardized work. These are described as follows:

1. *Cyclic*: Work that is meant to repeat at the end of the current cycle and continue with a regular frequency. This includes both Category A and Category B standardized work. Examples of this type of standardized work are seen in manual work cells with relatively short takt times. However, it basically describes standardized work that requires the person to repeat the same sequence of tasks continuously.

2. *Noncyclic*: Work that occurs randomly, or at least appears that way, and does not appear to repeat at the end of the current cycle, but that must follow the same work sequence when the work is required. This is typically Category C standardized work, but it can also include Category A and Category B when dealing with long-cycle processes. We will discuss long-cycle in a later section, but we use this term to represent a condition where takt time is longer than a normal work period (e.g., building two houses per year, etc.). Examples of Category C standardized work can include tending groups of machines, preparing a customer order at a fast-food restaurant, or performing tasks for long-cycle standardized work applications such as installing the plumbing in a new home. It generally describes tasks that must be repeated in work sequence, but are only required occasionally rather than repeating at the end of the current cycle.

We are going to focus first on the cyclic applications, as this is the simplest form of standardized work, and as stated earlier, a firm understanding of the basics is necessary if we are to expand the concepts into other areas. The application of standardized work to noncyclic tasks will be covered in a later section. But before we leave this section, there is one other important concept to consider. Not all jobs will be designed around advancing a single product or batch of products to completion. The worker may actually be operating multiple machines or performing multiple tasks that may cross several product value streams. There are many reasons for this, including better utilization of equipment or worker, specialized worker training that is in limited supply, low daily production volumes of individual products, etc.

Shigeo Shingo (1989, 156–161) explains the concept of a worker handling multiple machines. He says that there are two types of multiple machine handling: vertical and horizontal. The vertical type is called multi-process handling and refers to the operation of multiple machines that are of different processes. The horizontal type is called multi-machine handling and simply refers to operation of machines that are all of the same process. A good example of multi-process handling is a manufacturing cell. An example of multi-machine handling is a worker operating a bank or group of molding machines. In some industries, it is necessary to produce products in batches. Often this results because of machines with very long cycle times and it is necessary to have a worker handle or "tend" a number of machines simultaneously in order to better utilize the worker's time.

We will also see that in order to better utilize the worker's time, it is necessary to design the job so that the worker is not made to waste time by waiting on machines to finish, walking farther than necessary, applying unnecessary motions, and so on. However, it is not desirable to fill in the worker's time with unnecessary tasks simply to keep them busy. There are three basic issues that should be considered. Although they will be discussed further in later sections, they are listed below.

1. Tasks should always add value.
2. Tasks should be easy for the worker to recognize and execute.
3. Tasks should not hinder the flow of the product and thus need to be considered carefully.

One last point to consider is that efficient use of the worker's time is not always the most popular issue with many companies. Financial practices and expensive equipment and tooling can often lead companies to make decisions based on obtaining the highest utilization of assets. This may seem like the proper thing to do. After all, if we spend a lot of money on equipment and tooling, shouldn't we try to use them as much as possible? However, it is precisely this point of view that can easily lead to overproduction, one of the basic forms of waste and considered by many to be the worst because it can cause so many other forms of waste. This is why it is very important to ensure that the equipment and tooling supports the worker's standardized work and does not go beyond the minimum requirements to meet the customer needs. But in order to be able to determine whether the equipment supports the work as designed, it is important to be able to study the work as it actually happens. We do this by observing the situation, and based on our observations, we make determinations or gather more information. Because it is such an important concept, we want to discuss it in more detail before we continue with cyclic standardized work.

Chapter 2

Observation

Before we get too deep into *standardized work*, we need to discuss a very important aspect of its development. This aspect involves the ability to observe what is *really* occurring in the workplace or work environment as opposed to what is *supposed* to occur (or assumed to occur). It is easy to forget that just because something is intended to occur a specific way, we cannot assume that it actually happens that way in real life, especially with standardized work. Therefore, it is important to learn how to recognize and assess the steps that occur in the work process quickly and with a fair amount of accuracy. We start off by trying to better understand the concept of *observation* in general.

We observe by using our physical senses such as sight, feel, sound, and smell. Our brain then uses these senses as inputs, individually or in combination, to process the current series of events that are of interest. From this processing, we decide what the current event means to us both in learning and, if needed, in reaction. What events do you visually see around you when you are looking at specific moments in life? Do you know how to interpret what you saw and what you need to look for in your next observations to verify your interpretations? For example, these moments could be related to specific situations such as preparing for and then taking a trip to the local grocery store to purchase food.

If you could step outside of yourself and observe your own actions, what would you see? Would your actions always contribute *directly* to your trip? Are all the decisions leading up to a trip to the store totally obvious from visual observation only, or are some of the decisions based on past personal experience rather than current conditions? Does your thought process involve certain key items or tasks for your trip? What are the key items or tasks required to ensure a successful trip to the grocery store? How much detail will you need for determining these key items or tasks? You may be surprised at the level of detail that you may not have noticed upon casual observation of these events. As we will see throughout this section, observing involves asking ourselves many questions. These questions will be primarily based on visual inputs of repeatedly observed events or tasks. These types of questions will eventually lead you to address your given objective.

First, let us look at some of the questions we might have while observing the steps that may have been involved in preparing for your trip to the store:

1. Why did you plan to go to a particular store?
2. How did you decide how much food you needed to buy?
3. Did you prepare a list of items that were needed before you left?
4. Did you check the food you had on hand before leaving?
5. Did you consider the weather conditions?
6. Did you ensure that you had all the necessary items such as keys, money, coupons, and the correct attire for the weather?

The next questions about the steps that you could observe might involve your travel or transportation to the grocery store:

7. Did you take a particular route to the grocery store?
8. Did you have to adjust your original plans due to traffic delays?

Questions concerning steps that might be observed after arriving at the grocery store could include:

9. Do you have your list of items needed with you?
10. Were you able to obtain all the items on your list?
11. Do you have the correct amount of money required?

… and then finally, were you able to return home in a safe and timely manner? Let us review what we have just outlined in observation for preparing and traveling to the store. Consider Figure 2.1.

We have discussed steps that could be observed in preparing to go to the store, traveling to the store, and returning from the store afterwards. Each step consists of individual tasks that should be completed before the next step occurs. These steps can be interpreted in many ways, based on

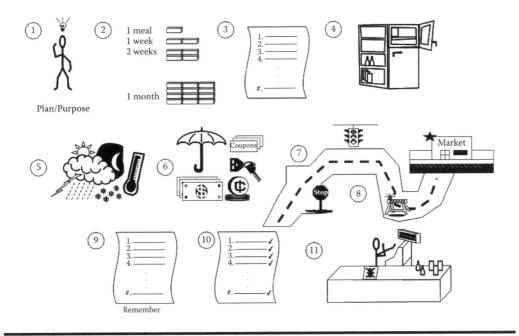

Figure 2.1 Steps observed during process of going to the store.

the perspective of the observer: Some tasks may seem very different, and some may seem very common. Each step could contain one or more tasks. Let us break down some of the steps into the corresponding tasks.

■ Go to the calendar—check date or day of the week
■ Check for special events on the calendar
■ Go to the desk—get notepad and pencil; get coupons
■ Go to food-storage shelves, refrigerator, and freezer—check food inventory and record
■ Go to car key area—get car keys
■ Go to window—check outside temperature and weather (is it raining, snowing, or cold?)
■ Go to closet—get coat and/or umbrella (if needed)

Steps 1 through 6 involved preparation for the trip. In observing this preparation, you could see an underlying or predefined thought process to try and minimize mistakes or potential causes for variation for the trip. Some of these mistakes could occur in the form of: (a) incorrect or omitted items needed from the store, (b) not having enough money to pay for all the items needed, (c) not having enough space to bring all of the items purchased from the store back home, or (d) improper attire for the weather conditions. We normally address questions to ourselves in a routine manner. We gain experience from the many trips we have already taken to the grocery store as well as other short trips around town. We also may recall experiences of others that we have observed or heard of, and some of these may also factor into our choices. Other times, we have no experience or information upon which to base a decision and just try different things and remember the results, otherwise known as trial and error. The overall effectiveness of your trip to the grocery store is usually determined by how well you considered the preparation in steps 1 through 6. We gain this understanding through the number of trips required to obtain the desired items—hopefully we accomplish this in a single, properly planned trip—as well as how well the trip went according to our expectations. The extent of this gain is based on the number of questions that you ask yourself each time and how well these questions relate to the main purpose or objective. This gain can also be developed through the observation of others conducting the same or similar tasks.

The following steps are an attempt to help summarize how you determine your criteria and the important or required steps into a common methodical approach.

1. Define the main purpose or objective for accomplishment in terms that can be easily understood by your target audience. (*Example*: Instructions on how to nicely cut the grass in your yard.)
2. Identify what you initially consider to be the major or key criteria required for success of the main objective. (*Example*: 1. Prepare Lawnmower.)
3. Identify the associated elements or steps of each criterion that you would consider important to the main purpose or objective. (*Example*: 1. Prepare Lawnmower: 1a. Check general condition of Lawnmower, such as wheel condition, blade condition, handle grip condition, and other safety devices; 1b. Add gasoline [as needed]; 1c. Add oil [as needed].)
4. Go to target area for observation, where the real work occurs, and reflect on your initial evaluation of key criteria and their associated elements. (*Example*: Go outside to your yard to check the condition of the grass. How high is the grass? Is the grass wet?)
5. Ask yourself, "Do my criteria and associated elements or steps align with my observations of where the work is actually to occur?"

(*Example*: Do I have my wheel height set correctly? Will the height of cutting leave clumps of grass or not cut the grass at all?)

6. If you find additional significant criteria or need to change your criteria and/or associated elements based on your observations, update your list from step 3, or else go to the next step. (*Example*: Add the adjustment of the wheel height step to the level of grass cutting that I desire for a nice-looking yard.)

7. Once you have verified your list of criteria and their associated elements at the work site, then you are ready to identify your read points for each element prior to your observational data gathering. (*Example*: We have added the wheel height adjustment step or element to the Prepare Lawnmower criteria based on observing the condition of the grass at the time of cutting.)

Making Observations and Formulating New Questions

Figure 2.2 provides a graphical look at the methodology used to set up the key criteria and associated elements for an archery example. The last step will take you to the establishing of read points and data gathering, which will be discussed next. The main objective in this case is to actually hit the target with the best possible outcome. The steps will walk you through an approach to be able to achieve the objective in a safe manner.

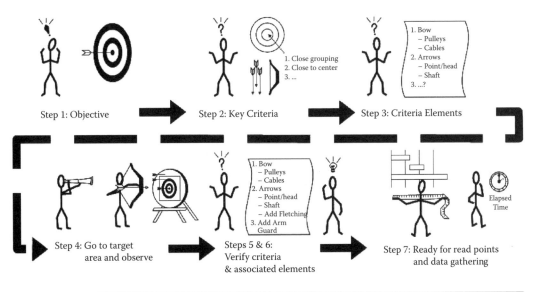

Figure 2.2 Method of setting key criteria for archery example.

Typically, when we first go to the workplace, we usually have a pre-conceived notion or expectation of what we will see. It is important that we do not let these expectations affect our interpretation of what we observe. We also will experience a certain level of unexpected observations. These unexpected observations usually become a new criterion or element for our purpose. The real key to making effective observations in the workplace or work environment is based on how well you understand the main purpose or objective; for us, this purpose is to observe what is *actually* happening. Most observations are biased by past experiences and memories, which can be greatly affected by our perceptions. The trial-and-error approach does not share the bias, but it is not a structured approach. Therefore, in order to improve on the trial-and-error approach, we define the important associated criteria and their complementary elements prior to our observations. This will allow us to learn more about what we observe in the field as we gather data—and it helps organize the approach by developing a focus for the observations. It is also very important to identify the criteria and elements that we want to observe prior to data gathering to improve the effectiveness and quality of the data. Next we will discuss how to break down our element observations in a format to support the study of *standardized work*.

How to Break Down Element Observations for Standardized Work

In our previous example, observation was discussed from a personal perspective by asking ourselves a series of questions. In reviewing this perspective, we tried to describe the process of observing. As we discovered, the best way to describe the process of observation is to look at the purpose or objective of why the observation is needed. In industry, the purpose of observation can be driven by many factors. Examples of these factors can vary from simple workplace improvements in 5S (Figure 2.3) to detailed time/extended time-oriented studies, such as the assembly of an aircraft or the transactional processing of a purchase order.

Consider for a moment that we plan to observe the process that a purchase order (PO) goes through at some small company. If we simply have the owner describe what is supposed to happen, would this be what actually happens? Suppose the process is described to us as comprising three steps:

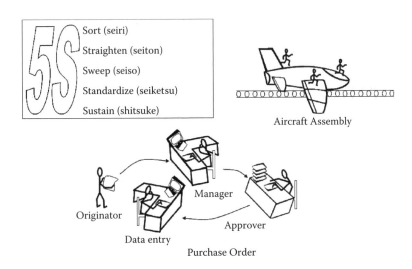

Figure 2.3 Example factors driving need for observation.

1. Originator writes a purchase order.
2. Finance approves the purchase order.
3. Data Entry enters the order into the system.

However, in this instance, when we go to where the work actually happens and observe the process, we notice that there are some things that occur a little differently, and there are some additional questions generated due to our initial observations. In this example, our initial observations are summarized in Figure 2.4. Notice that some of the additional questions are not explained well in the initial process description. If the instructions and details are not clear, then the workers of the system will have to decide for themselves the best course of action. This leads to variation between different people who follow this process, as shown in Figure 2.4.

The purpose or objective behind these studies usually has a common theme. This theme might involve the reduction of waste, understanding the sources of variation, the associated costs of the process, etc. So far, we have identified some items to help gain an understanding of how observation can be a powerful tool in our efforts to make improvements. If we use observation to look for and analyze sources of waste and variation in work elements from cycle to cycle within an operation or process, we also need a good method to capture the data observed. A common method that has been around for a long time is the traditional Industrial Engineering Time Study

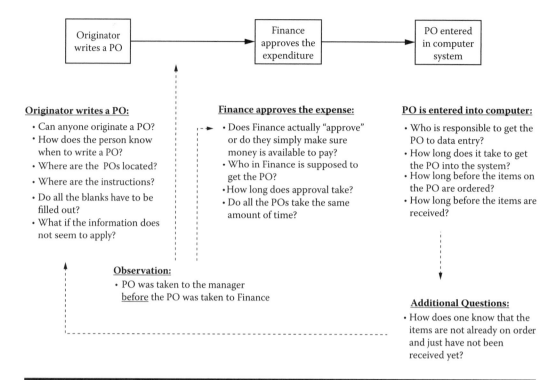

Figure 2.4 Observations of purchase order process.

element-breakdown approach. This approach looks at identified work elements in a modular or building-block form as a part of a complete operation or process cycle. The dividing line between work elements is usually established based on people-to-equipment or people-to-people transfer of work. The work elements can be further divided if finer analysis is required, but should be avoided for our purposes due to the complexity of data gathering and the introduction of human error. The following steps could be used to help visually identify work-element transitions within a process or operation:

1. Establish a defined Starting Read Point that describes the start of the work element observation cycle. This Starting Read Point should be a very easily identifiable cue for the observer to identify the start of the work-element task. This cue is often visual in nature, but can also be a sound. Examples include a person grasping a part or tool, loading a part, taking a step toward a work-element destination, a light or movement indicating the start of an equipment cycle, the sound of an air cylinder actuation, etc.

2. Determine the extent of this work element and the Ending Read Point by evaluating the transition potential from person to person or from

person to equipment. If no person-to-person or person-to-equipment transition occurs between the successive work elements, then the Ending Read Point can be established at the discretion of the observer as long as it can be observed easily. If it cannot be observed easily, then it may be best to combine the element with the next one in the sequence to reduce the chance for the introduction of variation in the observations and measurements.

3. Once the Ending Read Point (or simply End Point) for the work element is established, this establishes the Starting Read Point (or Start Point) for the next logical work element. This End Point should also be a very easily identifiable cue such as a sight or sound for the observer. Examples include a person releasing a part or tool, unloading a part, taking a step toward a work-element destination, a light or finishing movement indicating the finish of an equipment cycle, closing of a guard or gate, etc.

4. When the last logical work element has been identified and completed, then the End Point should match up with the very first work element's Start Point if the cycle repeats. This makes it very easy to take multiple measurements if you use a stopwatch with lap capability and multiple memories. The element data can be gathered quickly and efficiently if the read points are distinct. For more information on using a stopwatch with lap function and memories, see Appendix C.

Once the work elements have been observed, documented, and synchronized for their respective Start/End Points, then cycle-time data gathering can begin.

Establishing Work-Element Standards and Graphical Notations

A variety of paper documentation forms can be used to identify the work-element Read Points for cycle-time data gathering. We have chosen a very simple form (Figure 2.5) to use as an example, or you may choose a form that you feel fits your business better. Whatever form you decide to use needs to have the basic information as seen in Figure 2.5. The number of cycles you choose to gather data is usually driven by the main purpose or objective. It can also be determined statistically, by time availability, or resource availability.

Step #	Description	Read Points		Observation #									
		Starting	Ending	1	2	3	4	5	6	7	8	9	10

Figure 2.5 Example of a "read point" form.

If the preceding example were the subject of our observations, we would also find it helpful to use a simple graphical notation for describing the work elements as well as distinguishing one from another. We will learn more about this graphical method of notation in a later section, but we will briefly introduce it now in order to help our discussion of the read points. Referring to Figure 2.6, the solid horizontal lines represent actual human work that occurs at the location or the step number in time units. The wavy lines represent the actual walk time from one location to the other in time

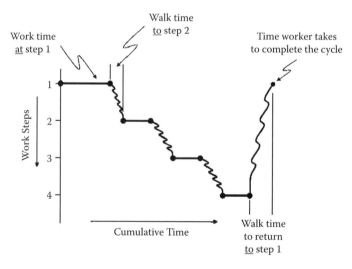

Figure 2.6 Graphical notation for describing work elements.

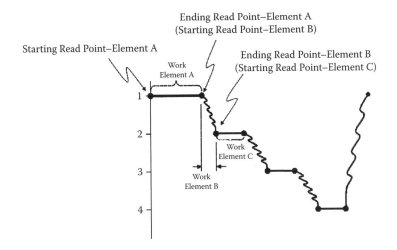

Figure 2.7   Example "read points" of work elements.

units, although the units are shown along the horizontal axis and not the
vertical axis.

Now if we apply the concept of read points to the graphic, we would
begin to see a relationship between the read points and the work elements,
as depicted in Figure 2.7.

We will learn more about this graphical technique in the section on
work-combination tables. However, it is important that we understand how
the various work elements are distinguished from one another and where
one ends and the next one begins in order to measure the individual work-
element times.

Methods for Work-Element Data

After you have captured the work-element data from your observations,
the next step is to evaluate your findings to help establish the baselines for
each work element of the person's job. There are many different methods to
normalize work-element time data from your observations. We will briefly
discuss two different methods for the purposes of this book: They are averaging and lowest repeatable. You may find other methods that are better for
your specific purposes or objectives.

The Averaging or Hi/Lo Averaging method for establishing work-element
cycle times is one of the more commonly used approaches. Depending
on the number of cycles and the maximum and minimum cycle-time data

points, high and low outliers may be disregarded from the work-element data set. The average cycle time for each work element is then calculated by adding up the individual sets of work-element times and then dividing by the number of individual work-element observations within each set. This is a very fast method to calculate the work-element average cycle time. Keep in mind that we use the word *average* cycle time. Even though high and low outliers may have been disregarded from your final average numbers, you are still calculating an "average" from a sample set of times, of which you have allowed variation to be factored into that average number. This can hide waste and sources of variation. The observed methods and workplace setup become an integral part of the observation data set for each work element. If you change either or both of these items, you may need to reobserve or restudy the entire person's job, depending on the variation interrelationships.

Another method that we have learned from Toyota is the lowest repeatable method. This method also looks within each set of work-element cycle times but uses a different approach. This approach takes each set of work-element cycle times and essentially lays the times in either ascending or descending order. The work-element target cycle time is actually chosen based on the lowest time that occurs most frequently, but must represent at least 25% to 30% of the total number of observations within the work-element set. The main point of this approach is twofold:

1. Establish a standard of time that has been demonstrated with some occurrence of regularity. The word *demonstrated* is stronger than just one occurrence, since this could represent an anomaly. It is important to keep the significant digits small so that there is reasonable distinction between the data points. So if you are using a stopwatch, you may want to round upwards quite a bit on the decimal places according to your needs.
2. Identify from the other observations within the work-element set the variations that cause those times to be different from the lowest repeatable time. This will draw more attention to the methods and workplace effects.

This methodical approach may require additional time to evaluate the variation difference within each work element. However, the advantage is to have a much better understanding of the cause and effect of the methods and workplace impact on the person performing the job. This will help reduce the amount of trial and error in continuous improvement activities and also allow the observer's efforts to be more productive. It is also important to note the need for easily observable start and end read points for the elements. If they are not very definitive, then variation can creep into your study and adversely affect your perceptions and interpretation of the data.

Evaluation of the Data

The chosen method should also look at the total time for each complete cycle from the first work element's Start Point to the last work element's End Point. You need to evaluate the total cycle-to-cycle variation that may occur. You do not want to effectively set up a job in a dependent work environment with mutually exclusive work elements. This can often happen with extended or variable equipment cycle times within a work-cell area. Taking a look at the overall total observed cycle time will put a sanity check on the capability of the standardized work. Suppose that Figure 2.8 shows a few typical observations and that 25 seconds was the lowest repeatable time observed.

As you can see from looking at the two different methods of calculating the work-step or work-element times, a glaring difference of approach shows itself. The Hi/Lo Averaging method allows for an extra 2.5 seconds of observed variation to be included in the overall average cycle time when

Work Steps	Observations						Method	
	1	2	3	4	5	6	(minus Hi/Lo) Average	Lowest Repeatable Observation Time
1	10	12	9	15	10	10	42/4 = 10.50	10
2	5	8	5	6	7	5	23/4 = 5.75	5
3	4	5	7	6	5	4	20/4 = 5.00	4
4	9	7	6	6	6	6	25/4 = 6.25	6
Total observed cycle time	28	32	27	32	28	25	Total = 27.5 (of averages)	Total = 25 (of lowest repeatable)

Figure 2.8 Example comparison of hi/lo and lowest repeatable methods.

compared to the lowest repeatable time. This amounts to an additional 10% of included time in the standardized work by design. Depending on the production requirements for the standardized work, you may waste a great deal of time in kaizen efforts dealing with the 2.5 seconds difference between average and lowest repeatable times. When you compare the Total Observed Cycle Times to the Hi/Lo Average cycle time, you will usually see that the Hi/Lo Average cycle time falls somewhere in the middle of the Total Observed Cycle Time sample, 25, 27, **27.5**, 28, 28, 32, 32. Once again, this shows the variation based on what the standardized work has demonstrated is actually possible. If we look at the Lowest Repeatable Observation times, we will see that we are accepting what the standardized work can produce at the lowest or quickest repeated level. The focus of our efforts should be to identify the variation contributor in each work step and work toward the improvements required to make the lowest repeatable method the norm. This is where the kaizen effort expended adds value to our original objective. However, it is also important to note that this method may never observe a cycle comprising all of the lowest repeated times for each element, and therefore the result represents a theoretical target time—not an actual time. This is why it is important to choose definite start and end read points that can be repeatedly observed to help eliminate introduction of variation by the observations made by the person gathering the data, such as reaction times with the stopwatch laps.

These are just two methods of determining a work-element cycle-time standard from a set of observations. Each method has its own advantages and disadvantages. You must decide for yourself which method will best fit your company's needs.

Please note that these types of observation techniques are of classic Industrial Engineering nature. Time Study is a classic version of observation that has been used in many industries and many fields of study. Some extra skills for the observer may be required to effectively measure these events. When looking at people performing the same tasks, it may be necessary, due to the nature or complexity of the tasks, to rate each person. This allows for the measure of variation between people. However, it is important to stress that the first step is to stabilize the situation. Once this occurs, the kaizen attitude should be used to start the process of transitioning to the desired condition. It is more important in the case of standardized work to stabilize the situation and begin the improvement process than to try to solve the problem by extensive data gathering and analysis.

The Importance of Observing the True Situation

One last note, as we try to observe, it is extremely important to remember that it is much easier to ask someone why they are doing something rather than try to determine this for ourselves by simply observing. Therefore, we should keep this in mind when we are observing someone else and not let convenience get in the way of our observations. What we mean by this is that, as you are observing a person going through a repeated series of tasks, there will often be reasons that are not always obvious. For example, if I envision myself watching you going through the work steps, as an outside observer I would not be privy to all of the thoughts and reasoning going through your decision-making process unless I asked you explicitly; I would know only what I could derive using my senses. However, one of the things we have learned is that we should not ask the worker why he or she is doing something without first giving quite a bit of time to our observations for several reasons. First, if we ask workers why they do something or why they do it a specific way, they may tell us something that is actually incorrect due to learning the method wrong, misunderstanding the original instructions, etc. We can also inadvertently influence the worker by the subject or nature of the questions. However, it is extremely important to talk with the people before making any extended observations in order to let them know what you are doing and the purpose of the observations. This is done out of respect for the person and common courtesy.

Second, if we interrupt them, they may actually do things out of their normal course of actions because they are being focused back on specifics, and it is important to observe things in the raw form—even if they daydream and forget steps, if that is what is actually happening. Third, if we cannot see the reason for a specific action, motion, etc., it causes us to intensify our observation actions to try and see what we may be missing. This is a good example of why some of our teachers would draw a chalk circle on the floor and tell us to stay within the circle, not talking to the worker or anyone else and just continue to observe. And finally, the observation activity should begin with no predetermined expectations, because often you will be surprised by what you *actually* see compared with what you *expected* to see. This is why we refer to the "current best method" used by the workers themselves rather than trying to surmise why the person was not making "rate" based on the "method they were told to use," which may be based on a preconceived notion.

Learning to See Below the Surface

One last issue before we continue with learning about standardized work is in what things to watch for in your observations. It is very easy to get distracted by the actual hand or machine motions when they are present. This is not to imply that they are not important, only that they are just a few of many aspects that we should get in the habit of watching as we learn to observe. For example, we definitely want to understand the motions involved in the hand work, obtaining of tools or materials, activation of the machines, etc. However, the person's hands are not the only points of interest. We need to watch their feet, their head, their torso, even their eyes—and if possible, it is also very helpful to observe more than one of the regular workers do the job. It is impossible to watch all of these things, so it is critical to observe many cycles to try and understand what is *actually* happening versus what is *intended* to happen.

There are many reasons for observing more than just the motions of the person's hands. The other points of interest can help us to discover things that were not obvious at first glance, which is one of the reasons for not interrupting workers to continually ask why they do something a particular way. However, it can also lead to discovering something that is not even evident to the person doing the work. An example of this can often be found by watching a person's feet while he or she performs the job. If you block the person from your view with your hand or a notepad so that you can only see the feet and the steps while the work is performed, you often find some interesting "dance steps." This is even more evident when you can observe more than one person do the same job. People with a smaller stride will need to take more steps than someone with a longer stride. The result of this often includes "half-steps," sidesteps, or other maneuvering of their feet in order to get into the proper position at each location. Noticing these "extra" motions of their feet can help lead to improvements and reduction of the variation between workers. The size of the equipment, the placement of the equipment or workstation, and other factors can be improved to reduce the variation caused by the differences in strides.

Another often-overlooked motion is that of the worker's eyes. Sometimes you can observe the motion of the worker's head which might lead us to notice the movement of the eyes. However, we can also make some other observations that lead us to their eye motions. For example, in most instances, the worker's eyes are normally focused where the hands are. This is just common sense, because not only is it the best way to do detailed work, it is very unsafe to be reaching in one direction and looking in another direction. If we observe intently, we can often notice this by not only their eyes, but also by their hands. If their eyes are looking in a different direction as they move their hands, we can sometimes notice this by hesitation of the hands (a short pause), or even a fumble of the motion (bumping the hands or something they are carrying into something along the way). Using this knowledge, we can look for some of the telltale signs of this condition. Suppose that the work at the location includes a machine check that provides a pass or fail indication. If the indicator is not in the direct line of sight with the worker's hands, then we must assume that their eyes are not going to move in conjunction with their hand movements. This causes the hesitation mentioned above, possible damage to the product by bumping it into an obstruction, or even dropping it outright. There are ways to measure the variation or delays caused by these situations once you learn to notice them in your observations.

Now that we have a better understanding of the importance of keen observation in the development and improvement of standardized work, we can now continue with our discussion of cyclic standardized work.

Chapter 3

Cyclic Standardized Work

As defined in Chapter 2, cyclic standardized work is work that repeats at the end of the current cycle and continues with a regular frequency. In other words, when the current cycle is complete, the next cycle is started, and the sequence continues to repeat. A normal work period, such as a shift, will contain one or more work cycles. An example of a single cycle might be a truck driver who makes a single round trip each 8-hour shift. But mostly, in order to develop a solid understanding of cyclic standardized work, we will be discussing situations where a worker is building or producing a product multiple times over a normal work period or shift. Assume for our discussion that the plan is for these cycles to align with takt time, as shown in Figure 3.1.

In this model, the worker is expected to produce many parts in the course of the work period. The cycle is repeated over and over until completion of the work or some disruption of the work cycle occurs. This disruption could be a problem, a scheduled lunchtime, the end of the work period, etc. Basically, the worker continues to repeat the work cycle as long as required, often for the entire shift. Consider an example in which the worker is building a toy truck, as depicted in Figure 3.2.

There are four basic steps to this work cycle:

1. Get chassis and engine and assemble together, then take to next station
2. Get body and assemble to chassis, then take to next station
3. Get decals and apply one to each side, then take to next station
4. Inspect truck for missing parts, pack good truck into box, then return to first station

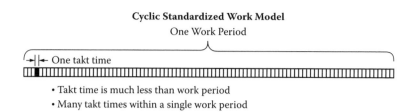

Figure 3.1 Cyclic standardized work model.

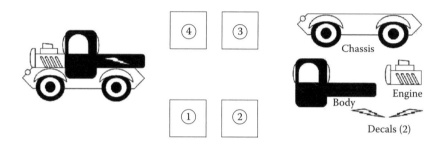

Figure 3.2 Toy truck workstations and parts.

Referring to Figure 3.2 and the work sequence listed above, we note that there are four basic workstations. The first station requires two materials, an engine and a chassis. The second station also requires two materials, the part that was assembled at the first station, often referred to as a subassembly, and a body. The third station requires three materials, the current subassembly from station two and the two decals (notice that there is a right and left version). Once the decals are applied, the assembled truck is taken to the fourth station, where the worker is to inspect it for missing parts and pack the truck into a box if it is good. At this point, the worker returns to the first station and repeats the sequence. We begin to see that there is more than just work involved, as we will discuss next.

Work-Component Types

In the previous example, we noticed at least two components that make up the person's work cycle that were obvious—working and walking. In fact, there are three basic components to consider. At a high level, each task comprises the following components (or their combinations) that a worker must do to complete the prescribed method and can be roughly classified as one of the following:

1. Work (motions or tasks that advance the product or service toward completion)
2. Walk (movement of the worker between successive workstations and materials)
3. Wait (idle or dwell time where the worker is forced to wait for a time or event)

During the work cycle, the product moves along on its journey to completion. Along this journey, the product changes in some way, has quality verified, is moved from one station to another, etc., all of which are things that may or may not add value from the customer's point of view. Upon further consideration, it is obvious that *work* is the most efficient use of the worker's time, as this directly adds value in some way. The next component is *walk*, and at this point the concept of whether the component adds value or not begins to get a little uncertain. An argument could be made that if a worker were walking to carry a part from one process to the next, then this contributes to the journey to completion. A fair question for this argument would be to ask ourselves whether the value to the customer increases if the distance walked increases. Some might argue that the worker is not adding any value, since the product is not changing or quality is not improving during the walk; although others might say that walking is necessary to move the product along as a means of transportation. It may not be required by the customer, but it is necessary at the moment under the current conditions or if other methods are not suitable. And finally, the last component is *waiting*. This refers to the worker being idle until a necessary condition allows continuation of the cycle. Waiting is one of the basic forms of waste.

Taiichi Ohno (1988, 57–58) explains these concepts very well. He basically says that, when breaking down the purpose and reasoning behind work and looking at the pieces that comprise the tasks, the pieces fall into one of three categories.

1. Work that adds value
2. Work that does not add value but is necessary at the moment
3. Work that adds no value and is not necessary

We can easily see that the last category can usually be eliminated straightaway because it is neither needed nor required. Often this is considered the low-hanging fruit, because it is usually the easiest to identify and eliminate. The second category, sometimes referred to as *incidental work*, is not always so easy to eliminate. Efforts to classify this type of work sometimes cause quite a bit of discussion, because it can be difficult to distinguish between adding value and being necessary at the moment. Therefore, we like the term *incidental* because it helps to identify the nature of the work. When looking at things from this perspective, it becomes easier to focus our efforts when resources are very limited.

The Effects of Variation

While analyzing the method for standardized work, it is necessary to consider the effect that variation can have upon the work cycle. The method will consist of various steps, each falling under one, or in some instances a combination, of the three basic work-component types. It is important to note that walking and waiting must be minimized in order to produce the parts in the most efficient and consistent manner. For example, how long does it take for a worker to walk 10 feet? Would all workers walk that distance in the same amount of time? What would the variation be from worker to worker? These questions help us realize that only by minimizing *walk* can the variation itself be minimized. *Wait*ing by itself only wastes the worker's time and does not add any value at all. However, we still must recognize that waiting may be necessary under the current conditions, but it is certainly a candidate for elimination in our kaizen efforts.

Try to imagine the effect of variation on each of the work-component types, and it is obvious that the time to complete the work itself can vary. It is also obvious that walking will produce variation between different workers that increases as the distance walked increases. But even waiting can vary if the worker is not assisted in some way. For example, a worker that is waiting on a machine cycle to *finish* is a very different situation than asking the worker to wait 7.5 seconds before moving on to the next step! The former has a definite ending that is observable by the person that can help create a "pace" signal, while the latter is very subjective without some aid to the worker. Because variation is one of the central targets of any improvement effort, it is important to recognize that it needs to be monitored along with the actual observations of the work cycle over time, as illustrated in Figure 3.3.

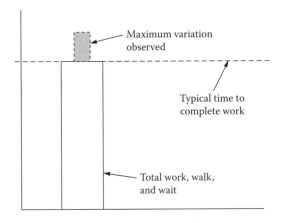

Figure 3.3 Example method of showing variation.

How to Document Standardized Work: The Standardized Work Chart

At this point, it is important to discuss how work can be documented. Making the work components more visual will help everyone have a better understanding of their impact. The standardized work should be summarized with a layout sketch. This sketch should show the overall work area, the work step locations, the sequence in which they are to occur, the walk path that the worker takes during the work cycle, and other pertinent information that we will learn more about as we go along. This layout is an important part of standardized work and we will refer to it as a standardized work chart (SWC), although it is known by other names as well. We will show more examples of the SWC later, but for now, it is important to understand that the first step in documenting standardized work is to understand how the layout affects the graphic depiction of the work.

For a work-cell example, we start by sketching simple representations of the workstations. It is not critical that the sketch be to scale or even that it be precise, although it is definitely helpful if it is at least a decent approximation. However, the order and general layout as well as the sequence are what we are most interested in depicting. Take the previous toy truck example of a work cell with four steps. Although you do not have to sketch it out precisely, it helps to show the correct relationship of the stations. Often we see a work cell laid out in a circular pattern in order to try and minimize walk (we will discuss this in greater depth later). If we were to try to sketch the cell, it might look similar to the illustration in Figure 3.4.

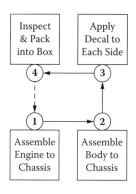

Basic Layout Sketch Rules:

- Sketch a rough approximation of each workstation and its location in relation to the other workstations

- Each major work sequence step is shown as a number within a circle

- The steps are connected by solid arrows

- The return to the step that begins to repeat the sequence is shown with a dashed arrow

- Work step, equipment name, etc. can be shown for clarity

Figure 3.4   Basic layout sketch rules.

The boxes represent the workstations. Notice that the work steps are shown as numbers in circles. The solid arrows are used to show the walk that occurs in the normal course of the work cycle as they move from one workstation to the next in sequence. The dashed line shows the walk that returns the worker to the starting point and signals that the cycle now begins to repeat. At each station, some amount of work is performed by the worker as well as a machine in some instances. In addition, the worker may be forced to wait for some length of time. If we are to properly understand the situation, we must be able to graphically describe the work components in some way. The first step is showing the geographic relationship between the workstations and the worker's movement between the stations as well as other pertinent information. The next step is to graphically depict the work components to serve as a basis for improving the standardized work. But first we will continue with our discussion of the standardized work chart.

We now know that the layout is a very important part of the SWC. We also know that the work sequence and worker walk path are important parts of the tool. But there is other information that is necessary or at least helpful to have on the SWC. You or your company may vary these as you see fit, but we want to discuss those that we have found to be necessary. One is the standard, or target, time that is needed in each cycle in order to meet the customer requirements. When determining takt time (TT), we notice that the calculation does not allow any room for losses or disruptions. This is a point of great disagreement in many companies. On the one hand are those that feel that TT is a target that cannot be met in reality, since there *will* be losses due to variation, defects, equipment problems, and so on. They feel that some allowance must be made for this in order to ensure that the requirements can be met over time. On the other hand are those

that feel that waste should not be designed into the system whenever possible and that using TT provides some "tension" in the system that provides
a pull for improvement efforts. Both ways of thinking have merit. There
could be a great deal of discussion here in analyzing the best method. For
example, in an industry that requires expensive equipment, it does not make
sense to purchase excess equipment "just in case." But when all the process
losses are factored in, the overall system losses can be fairly large in some
cases. Referring back to our toy truck example, suppose that each process
step had a 10% chance of downtime. This could be looked at as each process having a chance of running only 90% of the time. Since the processes
are sequential and there is no indication of in-process stock between the
workstations, if any one process stops, the result is the same as if all of
them stopped. The chance of no station being down at any particular time
(in other words the system is up and running) is found by multiplying the
run-time chance of all the stations together. In our example, the effect of a
10% downtime when applied to four processes in series is an uptime of only
about 65% (see Figure 3.5). It follows that the effect increases greatly as the
number of processes in the series increases.

However, in an industry that is labor intensive, it might make sense to
use the more aggressive TT as a target, since extra labor can be more easily
added or removed as needed. This is not the case in the expensive-equip-
ment scenario. It is important that your company decide for itself which way
works best for your situation, using TT directly or an alternative time that
allows for some losses. In this book we will consider the case where an
alternative time is used. The reason for this is that there are times when the
workload cannot balance perfectly when more than one worker is involved.
When balancing the work between workers, a target cycle is developed
based on the allocation of the work. We refer to this as the desired cycle
time (DCT), and it can be developed by considering more than just the customer requirements.

It is important to remember that if TT is used, there is no allowance for
losses for machine disruptions, defects, etc., and that the shortfall will need
to be made up by additional work time (most commonly this is through

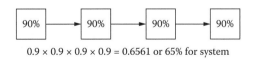

Figure 3.5 Cumulative downtime example for dependent processes.

overtime), at least until improvement efforts can eliminate the need for the extra work time. Consider that DCT indicates the desired cycle time, which implies that we have chosen a time that can take into consideration allowances for other problems such as machine disruptions, quality issues, etc., so that they can be tolerated without additional impact on the standardized work. Again, it is important to consider that building in allowances is a form of hidden waste and should be kept as visible as possible in order to continue to reduce the waste. Therefore, if you do this, we strongly suggest that you institute some form of tracking metric to keep it visible so that it can be used as a basis for kaizen. One idea is to track the ratio of DCT to TT (DCT ÷ TT × 100%). This ratio should be kept as close to 100% as possible. The lower it gets, the more waste is built into the system. Consider a system where DCT is 75 seconds and TT is 100 seconds. This yields a ratio of 75%, which means that we only need 75% of our resources to match customer demand when losses are excluded. In other words, we are carrying the other 25% of excess capacity for allowance of losses. If the losses are not 25% each day, then we will overproduce. But more importantly, how much waste is in the system *by design*? This is why many companies that have implemented TPS try to use a two-shift operation so that they can design much closer to TT and use daily overtime to make up for losses. Any daily overtime then is clearly visible and provides a driving force for kaizen activities. However, it can also be very disturbing if there is a lot of expensive capital equipment and tooling that is sitting around idle for a third of each work day. Again, this is a decision that should be based on your business and industry. One last issue to mention at this point is that of the practicality of the number chosen for use as a standard. It is important to consider that, when dealing with time in reference to human involvement, complexity can be introduced if the accuracy of the time chosen as the standard requires too much accuracy. For example, consider the difference between 35.4 s and 35 s. Is it practical to expect a human to be performing to standards that have less than a second of resolution? Often it is sufficient to simply round up to the next integer second.

We have defined and discussed DCT, but if we observe the work being performed by the workers, we do notice that inevitably there will be some variation from cycle to cycle when we try to measure things. In this context, we need to introduce a new term for discussion purposes that describes the time taken for a *particular* cycle, similar to a sample measurement. We refer to this as the observed cycle time (OCT), but sometimes we may also refer

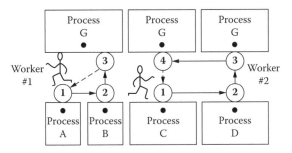

Figure 3.6 Two-person work-cell example.

to it as simply the worker's cycle time, since we know there will be some variation when comparing it to a standard time. While TT and DCT are standards, OCT is an actual time, like a sample observation. It will be useful for the analysis of problems later on as well as our current discussion, to which we will now return. Consider Figure 3.6.

In some instances where multiple people are working together in a flow system, not all the jobs can be balanced evenly. There may be a difference between the standard time for the entire system and the desired (or designed) time for each worker that cannot be eliminated easily. It is important to note that a person may be able to complete the work cycle more quickly than the preceding worker supplying the cycle and thus could be "forced" to wait for a few seconds at the end of the cycle until a part is available from the feeding operation. Whenever there are multiple workers in a system, handoff points should be used to decouple the workers when possible. There are two main reasons for this:

1. Ensure that the interfacing workers do not cross paths, which may cause interference with one another
2. Ensure that small variations between the individual cycles do not cause delays

These handoff points should contain no more than a single part, but this part will smooth out any mismatch in OCT that is one DCT or less. The discussion of buffers and replenishment rates in the section on worker job design will analyze this in more detail. Let us apply the handoff idea to the example of the multiple workers in the same cell. We see that it is necessary to capture this on the layouts, as shown in Figure 3.7.

While on the subject of waiting on a part, recall the earlier explanation of standard in-process stock (SIP). SIP differs from work-in-process (WIP) in

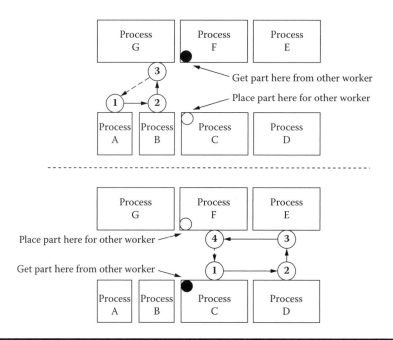

Figure 3.7   Handoff points for the two-person work-cell example.

that SIP is the stock that *must* be present in order for the cycle to continue to repeat as intended. For example, if the worker's standardized work calls for him to unload a previously completed part upon walking to the station, this implies that it is expected that a part will be there and be completed when he arrives. If not, a worker will not be able to complete the standardized work as designed. Therefore, it is important to show on the SWC what the correct SIP situation *should* be. This is done by drawing a filled-in circle (•) for single parts. (Notice that in Figure 3.7 the normal SIP that was present in Figure 3.6 has been removed in order to emphasize the handoff points.) It may also be necessary to show multiple parts at a station when required, and we can add something such as ×10 after the circle to indicate 10 parts. This symbol, when present, indicates that a part (or parts) must be present for the work cycle to repeat. It does not have to be a completed part if the worker does not require one (for example, if a new, unprocessed part were expected). SIP simply refers to what is required for the standardized work to repeat. The handoff points described previously should also be symbolized using the SIP circle or a variation, as shown in the table of Figure 3.8.

While on the subject of symbols, there are some other symbols that are commonly used on a SWC. For example, the diamond (◊) can be used to indicate where a quality check is performed, and a cross (✛) can be used to indicate a safety issue for the worker (not a safety product or safety issue for

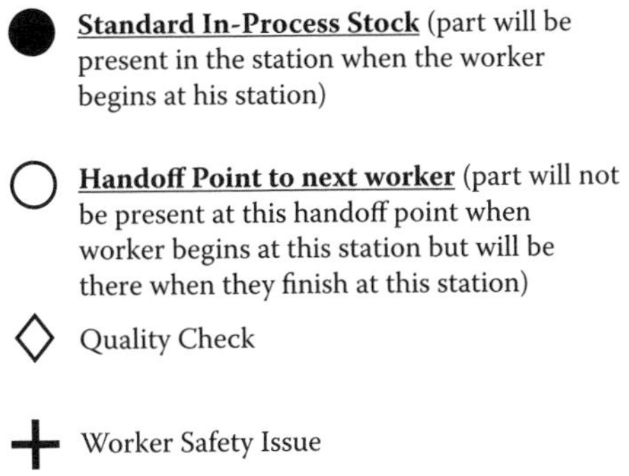

● **Standard In-Process Stock** (part will be
present in the station when the worker
begins at his station)

○ **Handoff Point to next worker** (part will not
be present at this handoff point when
worker begins at this station but will be
there when they finish at this station)

◇ Quality Check

╋ Worker Safety Issue

Figure 3.8   Example standardized work chart symbols.

the customer). If there are other important issues, such as a critical customer issue, many companies will add their own symbol to denote this. The
important thing to remember here is that all process issues are important,
but worker safety and product quality are critical. Think of the SWC as a
high-level summary of the worker job. It should be simple enough that it
can be shown on a single page while still showing the critical issues so that
it can be used for improvement (kaizen) or for aid in training new people.

There may be other pertinent information that is important in your application, such as job name/description, part number(s), variation/options,
revision dates, authorization signatures, etc., that may be necessary. Also, it
is common for machine or process names, cycle times, capital asset numbers (or other identification information) to be added to the layout sketch.
Another common practice is to cross-hatch the system constraint process
when present. When sketching the layout for the SWC, only the pertinent
equipment or geographic locations should be shown for clarity. Extraneous
information that is not necessary will only complicate the SWC. However,
it is important to note here that only one worker should be shown on an
SWC. The only exception to this is when more than one worker is required
to complete the standardized work as shown on the SWC. Otherwise, any
other workers involved should have their own SWC. Consider the example
SWC in Figure 3.9.

As we can see in Figure 3.9, the layout, work sequence, and walk path
are shown. The worker would expect SIP at each workstation, as denoted by
the solid dark circles. Also note that Process F has a quality check as well as
worker safety issues. But we see one other symbol used to denote handoff
points. Since not all systems have handoffs, it is often a good idea to define
them on the SWC. On the right side of this SWC, the symbol is defined for
handoff points to make sure there is no miscommunication about locations

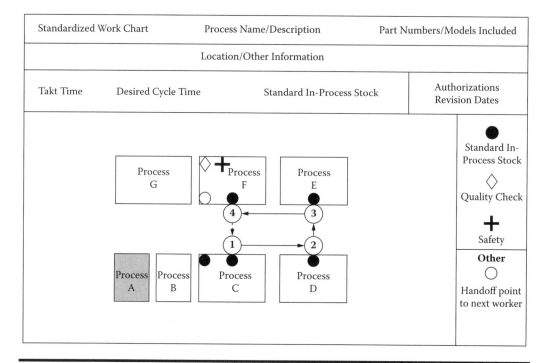

Figure 3.9   Standardized work chart example.

where the worker is supposed to exchange parts with another worker. We can see from the layout that there are other processes in the work cell, and therefore it is apparent that the worker is to receive the part for step 1 from another worker and is to pass the unloaded part from step 4 to another worker. These handoff points decouple the workers and make sure they do not cross paths, which would disrupt their cycles, as described previously. It is important to note here that a circle that is not filled in is being used for a handoff point that is supposed to be empty at the beginning of that task but will contain one at the end of the task, and is not used for a normal station that is intended to be empty—the latter being implied by the absence of a filled-in circle.

Handoff points are quite common whenever there are two or more workers in a cell. As mentioned previously, the handoff points allow the parts to be passed to another worker without having to walk into the work area of the other worker. The parts at the handoff points also allow for at least a single DCT to compensate for slight mismatches in the timing between the workers. However, it also affects the total SIP in the system, and therefore the number of parts at the handoff points should be kept as low as possible. In this example, we would expect a part to be present at the handoff point at step 1 when the worker arrived (thus adding 1 to the total SIP), but although there had been a part at the handoff point at step 4 when the worker left that station, we would expect that part to be gone when the worker arrived back at step 4 which is why we used the empty circle symbol. If the part was still there, the worker would have no space to place the part unloaded from step 4. If this were to occur, the worker would be unable to perform the standardized work as designed. One last issue to consider is that we should recall the diamond symbol indicating a quality check is performed at the same location as a handoff point. The diamond symbol is intended to remind us that care must be taken when dealing with parts that go through a quality check and make sure that parts are not mixed up so that a bad part is not inadvertently passed on to the next step.

Also note that Process A is cross-hatched, indicating that it is the system constraint. This means that it is very likely that there may be some wait time in this particular worker's job if the DCT is not closely matched with the other worker(s) in the cell. It should also be mentioned that the location of the system constraint, either before or after in the process flow, can have different effects upon the work. In this case, since it occurs before the beginning of this particular work sequence, it may cause a delay in the timely arrival of a part to fill the handoff point at Process C, causing the worker to wait before starting the next cycle. However, if the constraint process was located elsewhere in the process flow, the worker may complete a cycle before the part in the handoff point has been removed. This causes a significant problem because there is no "official" location for the pending part to be placed. In this situation, it is easy to simply place the part near the part already at the handoff point and continue. This is an example of overproduction and illustrates the need for a "pacing" method to be designed into the worker job. We will discuss worker job design at greater length in a later section.

One last point to discuss before we continue on is that these handoff points do cause additional motions that are not adding any value in the picking up and placement of parts. Consider that if a single worker was doing all the work, he or she would simply carry the part past the handoff point and place it directly into the subsequent process. Therefore, we must not forget that a handoff point is a compromise between the waste caused by the additional motions and the waste of the variation caused by the workers crossing into each other's pathways in instances where one operator is not sufficient to complete all the work tasks in the DCT. This compromise is meant to help reduce the overall variation in the standardized work so that kaizen can be more readily supported and result in the elimination of more waste.

Another purpose of the SWC, besides a basis for kaizen or training new workers, is for visual control. Often the same job is performed by more than one worker at different times. Different people may have different opinions as to what is the best method to perform the work. It is critical to remember the importance of everyone performing the work in the same manner to support small continuous improvements. The SWC can also be used as a tool for "auditing" the standardized work periodically to help remind everyone that this is the way that the work is intended to be performed. (This is discussed further in Appendix A.) However, it is also important to note that, in order to improve, we must not limit our kaizen efforts by thinking of the standardized work as ideal or permanent. But for the kaizen process to work properly, we must ensure that the situation is stable. The more stable the standardized work, the smaller the kaizen that can be attempted. It follows, then, that the smaller the kaizen that is attempted, the smaller the risk and potential cost impact (if the results are not as expected) that is assumed.

One last item to discuss while on the subject of the SWC is that of the complexity of the tasks or work steps. Some steps are so simple and straightforward that they need no other explanation. For example, consider our toy truck example. If the assembly of the engine-to-chassis and body-to-chassis tasks are simple enough, there is no other information or explanation required. But consider that step three has two decals to be applied. If there is additional information that is helpful to the worker at this step, then there is an additional tool that can be used. We will call this tool a Task Summary Sheet (TSS). Task summary sheets are very important for tasks that have tricks, tips, or other pertinent information for the worker. Although a TSS is useful for cyclic tasks, they are a critical issue when developing noncyclic standardized work. These will be described in more detail in the section on noncyclic standardized work.

Tools for Standardized Work: The Work-Combination Table

Now that we have a basic understanding of how to document the standardized work from a graphical summary point of view, we need to turn our discussion to the documentation and analysis of the work components. This is a more focused view of the work components that make up the summary

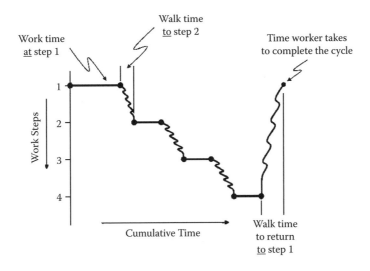

Figure 3.10 Simple work-combination table graphic.

shown in the SWC. Therefore, we need to visualize the individual steps in the work sequence and break them down into their work components. So when trying to understand the work, we begin to see that it is important to reduce the work to its basic components with no excess motions, as well as eliminate or minimize walking and waiting. This in turn becomes a balancing act between work that adds value and work that may not add direct value but is still necessary at the moment, and we need to have a firm understanding of how the three work components affect the standardized work. One of the ways we can better understand the effects is by developing a simple graphical representation. The most common graphical tool is called a work-combination table (WCT), which was briefly mentioned in the section on observation. A basic WCT for a manual job is shown in Figure 3.10.

In the basic work-combination table, the steps of the work sequence are shown along the *y*-axis beginning from top to bottom, as seen in Figure 3.10. Any work performed at a particular step is shown as a solid line moving in the *x*-axis direction from left to right, with the length of the solid line indicating the work time. Walk time from one step to the next is drawn as a wavy line, with the total distance in the *x*-direction denoting the actual time (regardless of the length in the *y*-direction). In the simple example given here, the worker performs some amount of work at step 1, then walks to step 2 and performs some additional work. The sequence continues this way through step 4, at which point the worker walks back to step 1.

Although there is much more than this to the work-combination table, it is important to understand the basic idea before we can truly appreciate the power of the other aspects of this visual tool. There are various printed forms of a work-combination table tool, but it is often sufficient to simply graph the standardized work in rough proportion on a blank sheet of paper—rather than using a preprinted form and taking care to ensure that each work component is drawn to scale—in order to illustrate a problem. It will be shown how this is possible shortly, but first we will discuss some more of the basics of the work-combination table.

In our WCT, after step 4 the worker returned to step 1. This implies that the worker can begin another cycle and continue the sequence. The accumulated distance along the *x*-axis from the point where the cycle starts to where it begins to repeat indicates the total time of the cycle. As defined earlier, this is the observed cycle time (OCT). If the worker is not consistent in the steps, the time between individual cycles can vary, sometimes considerably. Therefore, we begin to see why it is important to try to standardize the way that the work is done: It is so that we can observe the cycles of work, looking for abnormalities that cause variation so that they can be eliminated or reduced. However, how do we know if there is a problem with the OCT unless it is compared to a standard? For our discussions, we will assume that DCT is the standard (refer to Figure 3.11).

Previously, we indicated that it may be necessary to allow for various issues such as quality problems, machine downtime, or even the production of different products at the same stations. So it is important to have a standard to compare OCT to in order to determine the relationship between the work and the required rate. As we discussed previously, many companies

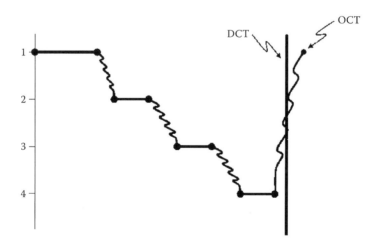

Figure 3.11   Comparison of observed cycle time to standard.

design the worker jobs to take into account some losses due to defects, machine downtime, etc. Therefore, instead of TT, they may have a number that is slightly less than TT to compensate. Again, as we discussed before, it is important to understand that this introduces waste into the system in the form of overproduction—which must be kept to a minimum. This is why we use DCT (*desired* cycle time), so that any waste that is planned into the work can be made visible in order to be a target for improvement. In our example, OCT is longer than DCT, so we can see that there is a problem because the work is taking longer than the time allotted. However, whether you use DCT or TT as the standard, if a standard is known, it should be shown on the WCT as a vertical solid red line. This is the standard time that we are trying to consistently meet each cycle. The comparison of the standard time to the OCT is also the basis of the audit philosophy that we mentioned earlier, and more information can be found in Appendix A. This standard is also an important part of the SWC that we described previously, and it will be included on the SWC as well.

Before continuing with the discussion of the WCT, notice that in the previous paragraph it was implied that there may be times when a standard time is not known. This is because the WCT can also be used for other tasks that are repeated but not necessarily cyclic. In a cyclic task, when the steps are done, the sequence and times are to be repeated each time. An example of a task that is repeated but not cyclic could be a lengthy changeover of a machine. The WCT can still be used as a tool for reducing the changeover, but in this instance a standard time is not as critical because the goal is to do the changeover in as short a time as possible (or at least less than the planned maximum time), whereas in a production work cell, the goal is to do the work in the desired amount of time every time until a better way is found and implemented in its place.

As the name says, the tool is a work-*combination* table. This means that it graphically shows the combination of all the work, for both the worker and any machines involved. It is common for a worker to be in a work cell that uses machines, and therefore these machines have work time as well. The WCT is meant to show the combination of the human work components (work, walk, and wait) as well as any machine time. So far, we have only shown human work components. To show the machine time in combination with the worker time, a horizontal dashed line is used to show the machine time for each step in the work sequence. Suppose our previous example contained a machine at step 1. The WCT might then look something like Figure 3.12.

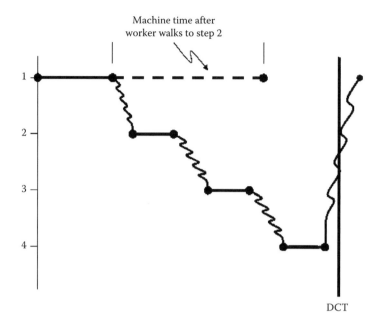

Figure 3.12   Documenting machine time on a work-combination table graphic.

It is also common for a work cell to have machines at multiple stations, depending on the industry or the product being built. If a machine cycle will cross the vertical DCT line (or TT if that is the standard—but applies to OCT as well), the machine-cycle line stops at the DCT, and the remainder of the cycle continues at the *y*-axis for that step, as shown in Figure 3.13. The reason for this is to help show whether the previous machine cycle will

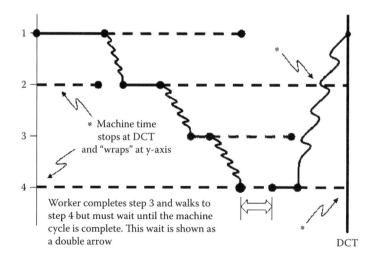

Figure 3.13   Documenting impact of machine time from previous cycle.

be complete before the worker returns to that particular step based on the standard time used. If it is not, the worker must wait on the machine cycle to finish before the worker's cycle can continue, as shown in Figure 3.13.

The Importance of Geographic Relationship

By now, we should be beginning to see that a simple WCT, when sketched properly, can show problems even when the scale or times are not very accurately drawn. This is because the y-axis does not have units of measure like the x-axis (recall that the x-axis is measured in units of time). So what is the purpose of the y-axis if there are no units? Its purpose is not to simply show work-step sequence, but to show the actual *geographic relationship* of the work sequence. This is important, since the purpose of a WCT is not just to document cyclic standardized work, but to serve as a basis for improving the currently best known method. If the problems are not accurately reflected, they cannot be properly analyzed for improvement. This is probably the most powerful aspect of the WCT as a tool for standardized work. The ability of the WCT to show a problem that may or may not be obvious when observed where the work actually happens can act as a common means of communication. Note that it can also show problems that are not obvious to an uninformed observer, such as the example where the OCT was longer than DCT, indicating that our standardized work was not meeting the standard. (Just because the DCT is the desired time does not guarantee that it is the typical time.) However, since the skilled worker can often make a series of tasks flow in a smooth sequence, this can often mask real problems and other forms of waste. Consider the previous example, where the worker arrives at a station only to be forced to wait for the machine to finish. In cases like this, it is common for the worker to slow the pace a bit to try to eliminate the waiting, which then covers the actual problem and amplifies variation between workers. Also note that if the WCT is not drawn correctly, some problems will not be documented and can only be observed by watching an actual work cycle. Consider the sketches in Figure 3.14.

In the first sketch, the work-step sequence and the direction that the worker moves coincide. This gives the WCT graph a stair-step like appearance. But in the second sketch, the WCT graph clearly shows that the worker movement is different. In this case, the difference is that the work sequence has been altered using the same layout. If the y-axis were shown only with the work-step sequence, then the second sketch would have

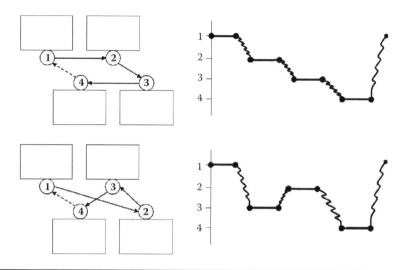

Figure 3.14   Documenting impact of process step location in relation to worker.

the same basic pattern as the first sketch, thus concealing a real problem. Therefore, we now understand that the *y*-axis is labeled showing the geographic location of the work steps in relation to the sequence. In the second sketch, we can see that the worker moves past the location of step 3 on the way to step 2 and then the worker moves past step 2 on the way to step 4. This last point—that the geographic location of the work steps is so important—illustrates why it is critical to develop these sketches while at the location where the work occurs. The standardized work and the tools associated with it are meant to be used where the work happens, not in an office or conference room. If you can observe the work actually happening, talk to those performing the work, and have tools that accurately reflect the problems and the situation, then progress toward improvement can be made.

Making the Problems Visible

Next, we will discuss the more common features of work-combination table paper forms. Normally, these include the step number, description of the step, the human and machine work times at that step, and the time to walk to the next step in the work sequence (not necessarily the next step line on the WCT). A WCT tool that allows the work-component times to be shown to scale can be very useful in the kaizen process. It allows each component

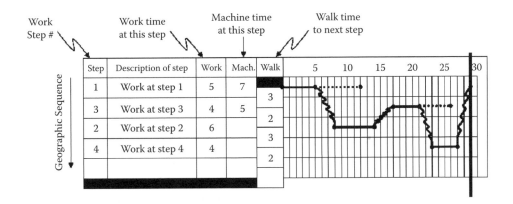

Figure 3.15   Work-combination table example.

to be quantified in relation to the others and can be used as the basis for improvement. There are endless variations of WCT forms, but most will have the following features in some form (refer to Figure 3.15).

We have already noted that the work sequence and the geographic sequence do not always coincide. As mentioned previously, it is critical to recognize that the walk time is the time to walk to the next step *in the sequence*, not necessarily the step in the next row. This is important to note, because the walk time between steps might not be accurately shown if we get them mixed up. When sketching a WCT on a blank sheet of paper, accuracy of scale might not be as important as it is when using a WCT form that is to scale. The required accuracy depends on the purpose of the WCT, so developing a layout sketch should always be the first step in creating a work-combination table so that any problems that are identified with the SWC or the layout are also visible on the WCT.

There are many problems that can be shown easily with a WCT. We will try to show several common problems and how they can be reflected using the WCT in Appendix B. But first, we should go back and discuss the SWC in more detail. We know that the layout of the work area and the geographic relationship of the various workstations are central parts of this tool. There are some other important details that should be included, and at this point we would like to talk about them and why they are important. There is no "best" way to draw a SWC. However, there are certain similarities between the various versions that we have seen. Again, it is not necessary to be extremely precise. In fact, we strongly believe that both the SWC and the WCT should be sketched by hand with pencil and paper. The reasons for this are simple:

1. They can be created and used right where the work happens, as opposed to doing them on a computer at someone's desk.
2. If they are drawn in pencil, they can be easily changed at the work site without needing access to a computer.

This would also allow the team members or the workers themselves to be a part of the kaizen process. This is such a strong belief for us as we write this that we are actually having quite a bit of difficulty creating our example illustrations. (A hand-drawn example loses a lot of clarity when scanned and inserted into an electronic document versus drawing examples with a computer program.) Ultimately, a clearer example can be drawn easily and quickly with a computer for the sake of expediency, but in the long run, the pencil-and-paper philosophy is what is required as part of the kaizen attitude. For more on how to document problems with a WCT, please refer to Appendix B.

Chapter 4

Long-Cycle Standardized Work

Up to this point, we have been discussing *cyclic standardized* work, where the work period is greater than the required customer demand or takt time (TT). We intend to show that, just because the worker is not producing many completed parts with very small takt times, the principles of standardized work are still applicable if we expand our thinking about the concepts. Before moving on to the application of standardized work principles to noncyclic tasks, it is first very important to understand the concept of long-cycle tasks. Compare the two diagrams in Figure 4.1. The first diagram is the same one that we used to introduce cyclic standardized work. However, notice that the only conceptual difference between the first diagram and the second is that the work period and takt time have been interchanged.

First of all, the difference is a subtle one, but it is extremely important to the concept of noncyclic standardized work. Most people would agree that it takes much longer than 60 seconds to build a single automobile. However, if you break up all the tasks that are required to build a car so that many tasks are happening in parallel, you can basically produce a car every 60 seconds. So even though it might take many total hours of work to build a single car, if it is possible to have enough resources and equipment committed to the purpose, it is possible to design a system to produce a car every 60 seconds. But what if, for some reason, we do not have an enormous amount of resources and equipment, or extremely high customer demand, or if the materials are extremely expensive or huge, or if the end product is not movable and therefore must be built at the customer's location? This means that there are many cases where the time between successively produced products will be much larger than our previous examples. Sometimes the

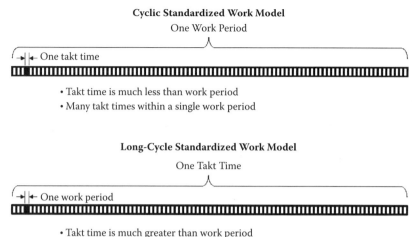

Figure 4.1 Comparison of cyclic and long-cycle standardized work models.

customer's demand is so low in such cases that the takt time calculation can result in a very large value—days, months, or even years! For the purpose of our discussion, we will normally refer to long-cycle standardized work as scenarios where the takt time is greater than the work period.

Let us focus on this last condition. Consider a company that builds houses. Disregarding for the moment things like prefabricated houses, how could the principles of standardized work be applied to building houses? If we compare the two illustrations above, we see that the only real difference is that the work period (shift) and the takt have been interchanged. In this case, takt time has become basically equal to the total time that it takes to build a single unit, in contrast to our automobile example, where takt time was the pace at which a completed product was output. If we look at things from this perspective, we start to see that this scenario can also describe other business situations, such as build-to-order or other similar circumstances where mass-production techniques do not apply well.

Applying Standardized Work Principles to Long-Cycle Applications

How can we apply standardized work principles to such situations? First, we must always remember to go back to the basics and consider the four preconditions for standardized work and make sure they apply. Recall that they are: (a) work that a human is capable of doing, (b) repeatable work sequences (we always build houses in a basic order, although some things can have their substeps rearranged somewhat for things like weather, etc.), (c) highly reliable equipment and workplace, and (d) high-quality materials. If these preconditions are all present, we have the prerequisites to support standardized work. Next, the three required components of standardized work—takt time, working sequence, and standard in-process stock—must be considered. We have already recognized that takt time in this situation has taken on a much longer time frame, but it is still there. The working sequence is fairly straightforward, since we have already determined that there *is* a repeatable work sequence as a precondition, although there may be some room for rearrangement (because some substeps can occur in parallel in many complex products or services) if conditions warrant, so it seems to apply. Finally, the concept of standard in-process (SIP) stock seems to apply very easily. In fact, since the same basic work crews must do most of the tasks, the SIP is probably one of the mitigating factors of the working sequence, along with the weather and other common random factors.

Although we cannot build a house by the strict rules we would apply to cyclic (maybe at this point we should say short-cycle rather than just cyclic) standardized work, there are some things that begin to show clearly for this example. One is that when a contractor's crew does a specific task such as install a wood floor, it is important that the crew does it basically the same every time. There are many reasons for this. One is for the amount of time allotted, because if they take too long, it can greatly affect the contractor's costs. Another is the quality, since the crew must have a wide variety of skills, and it is probably not feasible to have only experts do particular tasks. This would be similar to the automobile example described previously. If enough resources were available at the same time, it would be possible to complete a house in a much shorter time than a smaller contractor with limited resources. Also note that the former might not be the most cost-effective way to build houses. Therefore, the contractor needs to make sure that there is always something available for the crews to do that moves the house toward completion. If not, the contractor must assign them to work on something that moves another project toward completion; otherwise, the crew is being paid but adding no value, which is waste. This is good for the business, since it tries to make sure that the firm's crews are utilized properly, but it may not be so agreeable to the customer, since work on their house (product) may be delayed. This is why contractors must build allowances into their schedules for things like weather, material shortages, sick crew

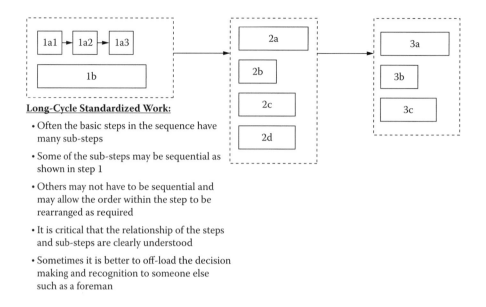

Figure 4.2   Example of long-cycle sequential and parallel work elements.

members, and so on, in their completion estimates to the customer. This is not unlike the earlier discussion about takt time (TT) and desired cycle time (DCT). Refer to Figure 4.2.

The Concept of Parallel Work Steps

Notice that we have introduced two new notational conventions to the concept of work-sequence documentation. The first is that of parallel tasks. Some tasks do not necessarily depend on sequence or may have some latitude on when they can occur during a portion of the sequence. They are still required, but simply need to be performed on demand. *These are denoted by letters rather than numbers.* The other concept is that there can be a combination of sequential and parallel tasks in the total work. *These are denoted by using numbers, letters, or a combination of both, as shown in Figure 4.2.* The reason for this distinction is that there are some tasks that are very complex. They may require additional workers (for example to maneuver large or heavy materials); they may require special training or expertise (an electrician or computer technician); or there can be many other reasons. Another important point is that knowledge of where a deviation in sequence can occur within a step may be very relevant to the proper balancing of the workers in order to maintain takt time or whatever standard is used.

Other Cyclic Standardized Work Issues

As we learned in Chapter 1, there are three basic categories of standardized work that are quite common. Categories A and B are both repeatable, but there can be differences between work content and total work times when comparing different products within a manufacturing system—and specifically within a particular worker job. By definition, Category B applications imply different times between the products run, and this can cause takt times that seem impossible to calculate without using some form of averaging, which may work for a particular volume mix but can vary greatly with changes in volume if the differences are significant. Also, when the work sequence is not repeatable (or at least it appears that way within a single work period), this is called Category C standardized work, and takt time does not seem to apply at all. These situations can cause great difficulty in considering the basics. We know that the preconditions still apply. However, when we consider the three required components, we often run into a problem with calculating takt time for Category A- and B-type applications. Consider the example in Figure 4.3.

Up to this point, we were considering takt times that easily translated to a (short) time per each product. For Category A standardized work, the number of parts required, often referred to as the *total volume*, was combined, and thus the takt time calculation was simply a reduction of the fraction (number of parts per time period) to its lowest terms. If the three different product symbols shown in the previous example represented batches of parts, Lean teachings would suggest that we try to build in the smallest

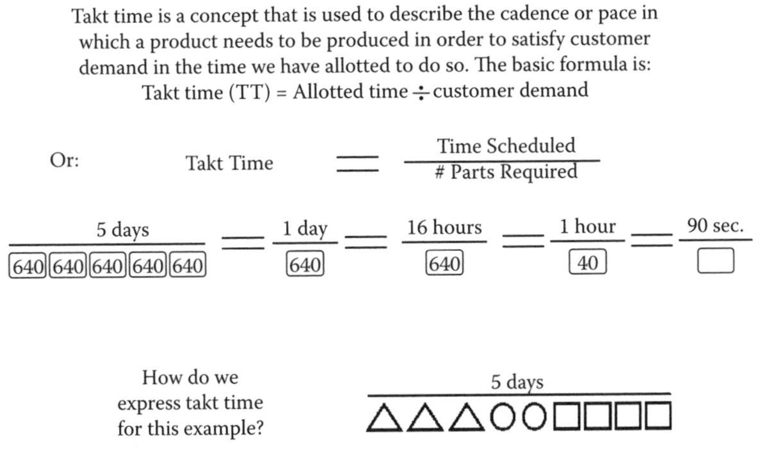

Takt time is a concept that is used to describe the cadence or pace in which a product needs to be produced in order to satisfy customer demand in the time we have allotted to do so. The basic formula is:

Takt time (TT) = Allotted time ÷ customer demand

$$\text{Or:} \qquad \text{Takt Time} \quad = \quad \frac{\text{Time Scheduled}}{\text{\# Parts Required}}$$

$$\frac{5 \text{ days}}{\boxed{640}\boxed{640}\boxed{640}\boxed{640}\boxed{640}} = \frac{1 \text{ day}}{\boxed{640}} = \frac{16 \text{ hours}}{\boxed{640}} = \frac{1 \text{ hour}}{\boxed{40}} = \frac{90 \text{ sec.}}{\Box}$$

How do we express takt time for this example?

5 days
△△△○○□□□□

Figure 4.3 Expressing takt time for non-similar cycle time products.

lot of each symbol in the order listed: in other words, a pattern of three triangles, two circles, and four squares, and then continuing repeating this pattern. But what if the fraction cannot be reduced easily, as shown in the given example? If each of the symbols represented a single part and the times were different between the products, the takt time calculation per a single symbol would be almost meaningless and certainly would be extremely confusing. Does this mean that takt time does not apply? No, it still applies, but it will be more difficult to express it in simple terms that make sense. The answer lies in again going back to the basics. We know how takt time is calculated, but what is the basic *purpose* of takt time? What was the original purpose that drove the need for an expression of this concept of takt time?

The Concept of Takt

Takt time is a representation of the rate at which our customers require parts and is based on the time we have scheduled to run the parts. It is used to determine whether we are producing to meet the customer's requirement according to how we allotted our time (are we ahead or behind?). In the case of Category A standardized work, it is relatively simple to compare the time elapsed and the total parts produced to see where we are according to customer requirements (or schedule), because the total of all the parts are combined as if the volume was for a single product type. This is because there is a direct, linear relationship between the time allotted and the products required during the period. In the case of Category B standardized work, it is more difficult to express this same concept. However, although difficult, it is not impossible. As described previously, the time between repetition of the pattern of three triangles, two circles, and four squares (at least in the lowest practical terms of this pattern) would essentially be takt time in the given example. When this concept is extended to Category C standardized work, the same principles apply, although not necessarily in numeric terms. The key lies in another Lean concept, namely that of visual control. Consider the seemingly nonreducible fraction of the previous illustration. How would we know if we were ahead of or behind where we should be? In this case, if at the end of the 5 days we had not produced three circles, two triangles, and four squares, we would not have met our requirements. But this would be too late to do anything about it, and we are forced to try to find a way to estimate the progress to standard (schedule) during the 5-day interval so that we could make the appropriate adjustments or institute countermeasures as required.

At this point, it is appropriate to mention that this problem is yet another reason behind the Lean concept of pursuing smaller lot sizes. In the Category B standardized work example, if the schedule could be converted to a smaller time period, for example 1 day rather than 5 days, it would be much easier to determine if we were ahead of or behind schedule. The smaller the scheduling interval, the more quickly we can get a sense of where we stand. With a 1-day schedule, we would know for sure at the end of a day (hopefully much sooner than that) whether or not we were on track. Therefore, the smaller the takt "pattern" that can be taken, the faster is the feedback. This deviates from the conventional thinking simply in terms of a time per individual unit, such as 30 seconds per piece, versus a pattern of mixed parts within the prescribed time period. If we go back and consider the origin of producing parts one at a time, we learn that Toyota had developed a philosophy of one-by-one confirmation (Kitano 1997). In other words, do not produce another part until it is confirmed that the previous part was consumed and was good. This basic concept is integrated into other Lean philosophies. One that immediately comes to mind is the concept of kanban. The kanban signal is our confirmation to produce and replenish, allowing fast confirmation of consumption and quality.

Looking deeper into the concept of takt, we begin to see that it is similar to a production schedule in many ways. It is like a part-to-part time schedule that needs to be maintained in order to meet the customer requirements. Thus we begin to see the importance of ensuring that the work being done is moving the product closer to completion by the planned time allotted and not simply by keeping the resources busy. Therefore, we should be working only on what is *needed now* instead of the attitude that it can be shipped (or sold) eventually. Most industries cannot survive with this attitude when trying to stay competitive.

Looking for Other Ways to Express the Concept of Takt

Getting back to trying to better understand the concept of takt time and why it is needed, it is imperative to be working on what is needed now (which implies *knowing* what is needed now). If production is getting behind, we then need to consider what immediate actions are required to get back on track. And if we are ahead, we should realize that this overproduction

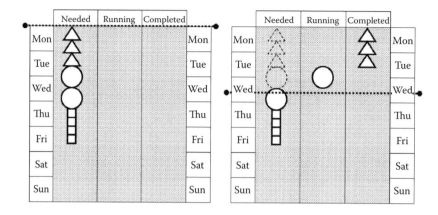

Figure 4.4   Visual representation of takt time concept for non-similar products.

increases inventory and thus be considering our options, since customers can change their minds, finished goods can be damaged if they sit around too long, and so forth. In fact, it is well known that Toyota considers the waste of inventory to be the greatest waste of all (Ohno 1988, 54). But again, how do we apply this concept to Category B standardized work applications? One answer is to see if we can find another way to express this concept instead of numerically.

In Figure 4.4, the requirements have been "scaled" to fit inside the time allotted. In addition, we have added some columns to show the status of the customer requirements. In this way, we can see if the particular order is waiting to be run, if it is in the process of being run, or if it has already been completed. If the horizontal dotted line moves according to the passage of time during the week, we could have a fairly accurate representation of the status during the scheduled period. The left table represents the start-up of the scheduled period. If the right table represents the condition at midweek, it appears that production is slightly behind schedule. This is because the horizontal dotted line, which crosses the symbols in the "needed" column, indicates which product should be running currently as compared to the original plan.

In these types of situations, we often see similar products grouped together for manufacturing to reduce the number of changeovers, especially in batch operations. Consider the situation when the build order of the products is changed considerably. The situation could look similar to Figure 4.5.

In this example, quick observation shows that we are right on schedule. In the first example, there were probably no more than three changeovers, but in this example, there could be as many as nine. It is obvious that quick

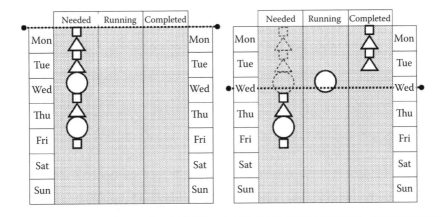

Figure 4.5 Reducing the batch size of previous example.

changeovers would be essential to this situation, where the original volume mix was leveled somewhat. However, we begin to see another familiar Lean philosophy arise—that of small lot production (or even one-piece flow, which is the case here). It is not our intention to discuss the merits of *heijunka* (or leveling the production schedule), but only to note the interrelationship with other Lean principles. Also consider that, in this example, circles and squares would be available to the customer in smaller quantities but much sooner than the first example. This can offer a competitive advantage in many industries because producers may not need to wait until an entire batch has been completed before they can ship (and get paid).

It is probably a good idea to discuss further the concept of quick changeover times between different products. It makes perfect sense to try and reduce the time for a changeover because changeover time adds no value to the product. However, the real issue is what the newly saved time from the reduction in changeover time is used for. If the time is simply used for extra production time, this allows us to overproduce, especially if we have a system that has longer changeover times factored into it. The real purpose behind this drive is to allow us the opportunity to do more changeovers than before. The greater the reduction of the changeover time, the more changeovers will be possible overall. This helps support small-lot production and even one-piece flow. As the changeover time shrinks and approaches zero, the lot size can become smaller and smaller until the lot size is 1. However, it is important to understand that changeover time between different products or services is still waste (affecting our costs) and should be a focus of our improvement efforts if we have the kaizen attitude.

Where might a situation occur in real life where excessive changeover time would greatly affect profitability? Consider a restaurant where the customers are served dinners from a predefined menu. Customers can order whatever is on the menu, and are free to change their mind up to some point. But the restaurant cannot accommodate too many changes by the customers without it greatly affecting the profitability. For example, if a customer changes his or her mind on a seafood dinner, it may not always be feasible to save this dinner for another customer, since you cannot guarantee that it will be required within a reasonable amount of time. It also is not feasible to have all the lobster dinners made up in advance (or stocked up in finished goods for reheating at a later date—probably not a good idea). Since the restaurant must be able to prepare any dinner on the menu within a reasonable amount of time, it is imperative that changeover from one entrée to another is kept to a minimum; otherwise, costs go up or health concerns may arise, and this could be bad for both the restaurant and the customers. This situation is not unlike production in many manufacturing facilities. But if customers want to be able to change their minds, or wait until the last minute to place an order, we must be able to accommodate this to a certain degree if we are to survive in a competitive environment. So we can begin to see the advantages to being able to respond quickly to changes in demand and how quick changeover supports this competitive advantage. The more quickly the product is delivered, the less time the customers have to change their minds (and the more quickly you can bill them or get paid for the goods or services). So, if changeovers are critical to your business, then applying the standardized work chart (SWC) and work-combination table (WCT) tools are essential to support improvements.

Our discussion on the concept of takt—or the thinking behind why the concept was needed for the short-cycle applications of standardized work—has shown us that it is essential to know how we are doing in relation to what we should have done in the same time frame. In other words, we need to be working on the right things at the right time, which means knowing what should be done and when it should be done. The concept of takt works similar to a schedule, except that there is a definite order to the total schedule rather than simply a list of totals that must be filled within a certain time period.

Chapter 5

Job Design for the Worker: Understanding the Levels of Interfacing

Now consider work involving manufacturing processes that are very time intensive or long machine cycles that do not require constant worker attention. Examples include machining processes, integrated-circuit-wafer fabrication processes, complex robotic welding or assembly, etc. In this type of scenario, the worker may only be required at periodic intervals of the total process cycle. When worker attention is only required intermittently, it is common practice to design the worker's job so that other work is available to better utilize the worker's time, such as operating multiple machines or multiple processes concurrently. We will discuss later how standardized work can be applied to tending machines with longer process cycle times if the worker job is designed correctly. But how are we supposed to design a worker's job when it is not straightforward and cyclic in the same sense as a work cell? First, it is important to understand the different ways that a worker may interface with processes involved in accomplishing the work.

In Figure 5.1, we see that there are three basic levels that a worker would interface with the processes. The 1st-order level is where the worker contributes to directly advancing the product through the production process. This could be by simple loading, unloading, and moving a single part (or even a container of parts) between the processes, or it could mean performing work on the parts, operating the controls of the machine, etc. The 2nd-order level is where the worker performs tasks that are required to directly

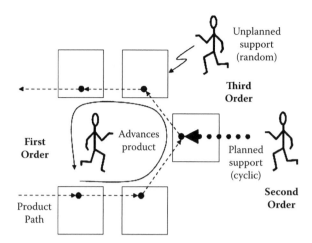

Unplanned support (random)

Third Order

First Order

Advances product

Product Path

Planned support (cyclic)

Second Order

- **Level 1** – First Order interface: tasks that directly advance the parts through the production process (not necessarily at takt)

- **Level 2** – Second Order interface: tasks that are required to directly support level one such as material loading, quality checks, change totes, etc. (predictable – based on multiples of takt, number of parts run, elapsed time, etc.)

- **Level 3** – Third Order interface: tasks that are related to random events such as machine repair, adjustment, etc. (random in frequency)

Figure 5.1 Basic levels of worker interface with processes.

support the 1st-order level, such as material stocking, performing quality checks, exchanging containers, etc. These are normally predictable in some way—based on multiples of added components, amount of material added, number of parts run, or elapsed time, but do not directly advance the parts toward completion. The 3rd-order level is where tasks that are related to random events, such as machine repair, adjustment, etc., are required to support the 1st-order level but are not typically predictable. It could be said that everything else in the business that is not one of these three levels might be considered the 4th-order level. These would consist of tasks that are shared by multiple products, such as Purchasing, or even things like clearing the snow off the parking lot. However, for the purpose of job design, we are only concerned with tasks of the first three order levels, which have the most direct impact on product flow.

Decoupling—When Is Protection from Interruption Needed?

It is important to understand exactly how the three main interface levels affect the production of the product or service. For example, consider what happens when the worker is interrupted or delayed at each of the levels. If this occurs at the 1st-order level, the product is in jeopardy of not completing normally if the worker's time is fully utilized. This could simply cause a slight delay in the completion time, or it could introduce the opportunity for an error to occur, or it might even mean that a defect is created (or a batch of defects, depending on the process). For the 2nd-order level, it might cause any of the problems listed for the 1st order, possibly to a greater degree, so how do we make sure this does not happen without having a dedicated worker standing there waiting for a signal to perform the necessary tasks? The most common way is to create a margin of "safety" by decoupling *their* work from the 1st-order work. For example, this could be done for the tasks involved with maintaining materials by creating a buffer. This buffer would offer some amount of protection from delay in replenishing the materials for the 1st-order-level tasks and allow more efficient time utilization of the worker responsible for the 2nd-order-level tasks. However, problems can arise if this buffer is not maintained correctly. Therefore, a brief discussion on buffers may be helpful.

Protection Expressed by Units of Time

In Figure 5.2, there are four types of buffers. In this context, we define a buffer in the same way as the theory of constraints (TOC): that buffers are measured in the amount of time offered for protection from disruptions (Goldratt 1990, 124). Although there is SIP (standard in-process stock) as well as normal work-in-process (WIP) in the diagram, the buffers are locations where extra material or WIP is *planned* for protection from disruption in the flow of the product through the system. The first buffer we observe is labeled Buffer 1. It is protecting the constraint process from starvation due to disruptions in the preceding processes. As we have learned from TOC, the

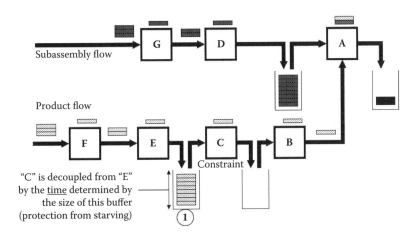

Figure 5.2   Example of buffer protecting the constraint from being starved.

constraint is what prevents the system from producing more output, and at any one time there can be only one constraint, although there may be several processes that are very close to being the constraint. If we analyze this first buffer and its purpose, it becomes obvious that the larger this buffer, the longer the constraint is protected from disruptions from processes E and F as well as stock-outs of the raw material at process F.

It is important to note that, in order to ensure that this buffer is kept at a practical level, it should be sized so that it is as small as possible while still covering a reasonable percentage of the disruptions. This is not as difficult a task as it appears. The buffer size can quickly be set by trial and error. First choose a time, for example 10 minutes (or whatever seems appropriate for the system), and monitor the buffer closely. If it runs out of parts too often, increase the size and continue monitoring. If it seems to never drop below a certain level of parts, reduce the size and continue monitoring. The most critical point to be made here is that, since Process C is the constraint, by definition Processes E and F can outproduce it and have extra capacity for replenishing the buffer. Therefore, they must not be allowed to produce more than what is needed to maintain the constraint buffer.

The next buffer we observe in Figure 5.3 is labeled Buffer 2 and is also intended to protect the constraint, but this time it is from blockage by disruptions in the subsequent processes. Because the constraint determines the output of the system in our example, any time lost at the constraint cannot be recovered unless more time is added to the system. This is because the constraint is internal to the system. If we analyze this second buffer and

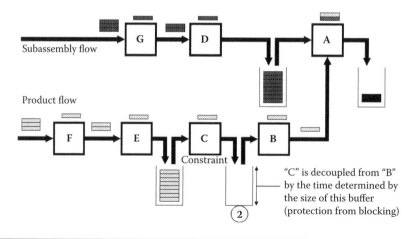

Figure 5.3 Example of buffer protecting the constraint from being blocked.

its purpose, it becomes obvious that, under normal conditions, we should observe no material in this buffer at all unless Process B is stopped. Its purpose is very narrow, and that is to protect the constraint from disruptions at Processes A and B. For most applications, this buffer is simply a temporary location to allow Process C to continue for some amount of time to prevent it from being stopped by short-term disruptions. Again, this type of buffer also requires close management, because allowing Process C to continue is overproducing, and therefore must be kept to a minimum. We describe it here only for the sake of clarity and for systems that have internal constraints that can prevent the customer demand from being met.

The next buffer we observe in Figure 5.4 is labeled Buffer 3, which is protecting Process A from disruptions from Processes D and G so that parts in the system that have already been through the constraint process are not delayed. We know that the constraint process limits the output of the entire system, so from a certain point of view any parts that have been through the constraint already are more precious than material that has not been through the constraint yet or does not go through the constraint process at all.

In Figure 5.5, the last buffer we observe is labeled Buffer 4 and is a finished-goods buffer that protects the system from variations in customer demand. This type of buffer is probably the most widely recognized type of protection and needs no further explanation. Although there are other forms of buffers, these are sufficient for our purposes—that buffers are used for protection against disruptions for a planned margin of time.

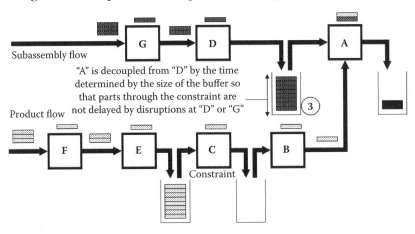

Figure 5.4   Example of buffer protecting constraint parts from subassembly problems.

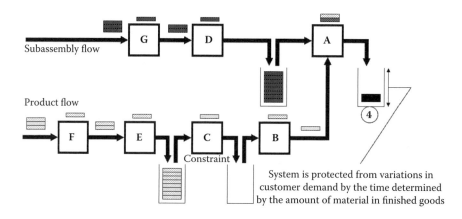

Figure 5.5 Example of a buffer protecting system from demand variation.

Effects of Coupled Jobs

The protection is simple: If there are the equivalent of 5 minutes of parts in a buffer, this provides a 5-minute safety margin before a disruption affects the protected process. However, there are several issues with buffers that require attention and management. Two major issues are the replenishment and consumption rates. If the rate of consumption is faster than the rate of replenishment, the buffer will be empty most of the time, like Buffer 2 in the previous example. If the replenishment rate is faster than the consumption rate, then the buffer will eventually fill up, and the preceding processes will need to *stop producing* until the buffer needs parts again (one of the many purposes of *heijunka*, or leveling the production schedule). For the purposes of our discussion concerning the 2nd-order tasks as defined previously, we must make sure that replenishment of material takes place *before* a disruption occurs. Therefore, the job design of the person responsible for this task has to be such that there is sufficient time to do the other tasks and still have time to replenish stock, move containers, etc. This can be a problem if the worker has a large area of coverage to maintain. Standardized work for these types of jobs will be discussed in the next section. But in our current discussion, we need to understand that work at the 2nd-order level of interface should not disrupt the 1st-order tasks; otherwise, product flow stops.

This concept is similar to mismatches in the work balance between multiple workers in a cell. Notice that in Figure 5.6, the desired cycle time (DCT) is set by worker 1. This is a situation that often occurs to various degrees when the work cannot be divided equally among the workers. The handoff

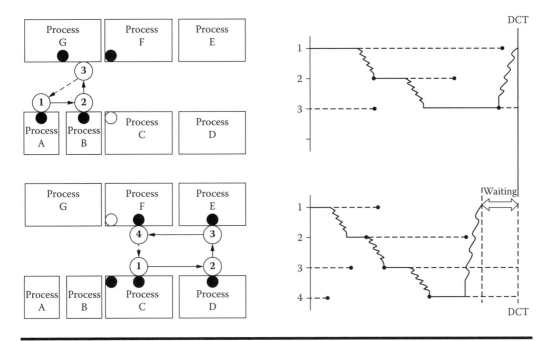

Figure 5.6   Impact of coupled jobs in a work cell.

points allow a buffer of one part to decouple the workers. The amount of protection for this example is one standard time (DCT). It should also be evident that the two parts at the filled-in circles indicating the handoff points are required to make this system work as designed between the two workers, which unfortunately includes a forced wait to keep the two worker cycles in balance. The wait time at the beginning of the second worker's cycle is necessary in order to keep things synchronized at the proper stop points. (The normal SIP required for these jobs has been removed to better highlight the worker interface points.)

As we have seen in our brief discussion of buffers, this is one of the ways that disruptions can be avoided for short delays when things are planned out or predictable. But there are other tasks at the 2nd-order level that can cause disruptions: material delivery, expendable tool changes, quality checks, process monitoring activities, and so on. The approach to these types of disruptions is to try to minimize them in a similar manner as minimizing changeover times, e.g., by identifying as many steps that can be externalized as possible, having a replacement part when a part must be removed from the line, etc. However, it is important to note that in the example considered here, the wait time is forced at the end of the cycle because of the mismatch in work balance. If the observed cycle time (OCT) for worker 1 was

28 seconds and the OCT (without the wait) for worker 2 was 26 seconds, the wait time would be 2 seconds. Because the replenishment rate (28 seconds) is greater than the consumption rate (26 seconds), the 2 seconds of wait is forced because there will only be a part available from Process B every 28 seconds. Also note that it can be an easy temptation to adjust the walk pace slightly in order to not have to wait at the machine for the 2 seconds. We have already discussed that it is difficult to judge time accurately, and it would be easy to misjudge the pace and begin to add variability back into the time for worker 2. It is therefore obvious this is not a desirable condition and is definitely a candidate for kaizen. However, it should also be noted that the job is designed such that the forced wait is natural. The problems begin to arise as the forced wait becomes obscured or absorbed by adjusting the walk pace such that the normal condition (a forced-wait problem) is no longer visible. Consider that if the times were reversed, in other words the first worker time was 26 seconds and the second worker time was 28 seconds, there would be the potential to overproduce if worker 1 has access to additional material. If this were the case, the forced wait would not be natural but, rather, somewhat "voluntary" to worker 1 and could cause even more problems.

Normally, 3rd-order tasks occur at random intervals. They include tasks such as process adjustments, equipment repair, and so forth. Planned preventive maintenance (PM) that is on a schedule could fall under the definition of 2nd-order-level tasks. If the downtime for the PM task is predictable and is performed on a regular basis, it would be considered a 2nd-order task; otherwise, it would be considered a 3rd-order task. Because 3rd-order tasks are normally unpredictable in occurrence and duration, these randomly occurring tasks must be minimized as well if the flow of the product is to be protected. These 3rd-order tasks often cause the longest disruptions in flow, and therefore this type of work must be very responsive in order to minimize the impact.

Using the Interface Levels for Design of Good Standardized Work

The reason that we have defined these interface levels of tasks is that they are relevant to job design and critical to the development of *good* standardized work. It is the framework for applying standardized work principles to

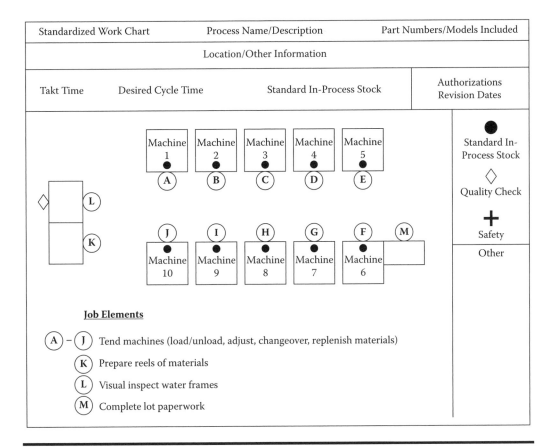

Standardized Work Chart	Process Name/Description	Part Numbers/Models Included
	Location/Other Information	

Takt Time	Desired Cycle Time	Standard In-Process Stock	Authorizations Revision Dates

Job Elements

A – J Tend machines (load/unload, adjust, changeover, replenish materials)

K Prepare reels of materials

L Visual inspect water frames

M Complete lot paperwork

Figure 5.7 Example SWC for worker job tending ten machines in parallel.

noncyclic work. For example, as we design standardized work we must be aware of the levels of the tasks and the impact that one has upon the other. Consider a multiple-machine example where a worker job is designed to operate 10 identical machines concurrently. Because the machines are running in parallel, there is no dependency of sequence of the load and unload order of the 10 machines. If this is the case, the standardized work chart (SWC) might appear similar to the illustration in Figure 5.7.

However, if we consider the dependency of the sequence of work between the machines, it becomes apparent that the worker could be moving from one machine to another in a more or less random manner. If this were the case, the walk time to the next location in the sequence would cause enormous variation in the work cycle. If the 10 machines were running different parts with different cycle times, this might indeed be the case. But suppose the worker was handling 10 identical machines running 10 identical parts. The situation now changes, and the time at which a machine

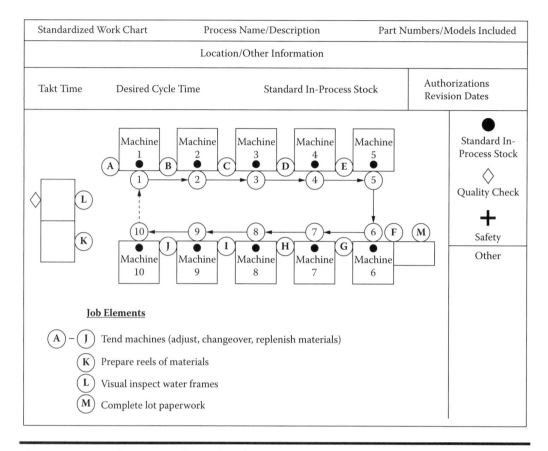

Figure 5.8 Previous example updated to reflect better worker job design.

will need to be tended to becomes very predictable. Although the order in which the machines are tended to by the worker is not explicitly required to be in a particular sequence, it is now very desirable to standardize the sequence to establish an order that can be repeated to reduce variation and stabilize the situation. Now the SWC might look slightly different, as shown in Figure 5.8.

In this situation, the load and unload portion can be standardized, and the variation can be better controlled. Although the work was somewhat "standardized" in the beginning of this example, the variation would have been enormous if the worker was forced to constantly look around to find where to go to for the next step in the sequence. So even though the original situation was marginally stable, it was far from ideal, and therefore we can begin to see that there can be situations where the work is standardized but not very good. At this point, it is important to note that in some situations, the first version of the example may be as good as we can get if the cycle times between the machines vary greatly. But if we suppose for the moment that all 10 of the machines are indeed running the same product, then we see that it is possible to make further refinements to strive for better standardized work. Now, under these conditions, we will look at some of the other issues surrounding the handling of multiple machines.

Returning to the updated example, under normal conditions, the worker is supposed to unload a finished part, load a new part, start the machine, and move on to the next machine and repeat the same steps (note that the worker does not transport a part between machines). In this example, each machine has a process time of 370 seconds, 20 seconds of load/unload time, and 20 seconds of walk time to reach the next machine. At each machine, the worker must perform 20 seconds of work then walk 20 seconds to the next location to repeat the cycle (40 seconds total). This is repeated 10 times for a total of 400 seconds, which we will establish as the desired cycle time (DCT). The total machine time that is needed to complete a cycle at a machine is 390 seconds (20 seconds load/unload plus 370 seconds machine cycle). Figure 5.9 reveals that each machine will be idle waiting on the worker for 10 seconds (400 − 390 seconds). Also note that any SIP has been removed for the sake of simplicity.

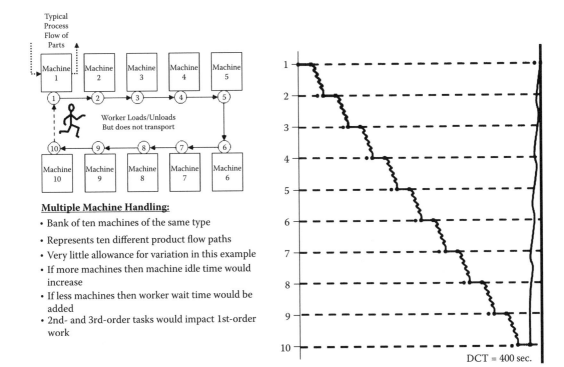

Figure 5.9   Worker job design issues for ten parallel machines scenario.

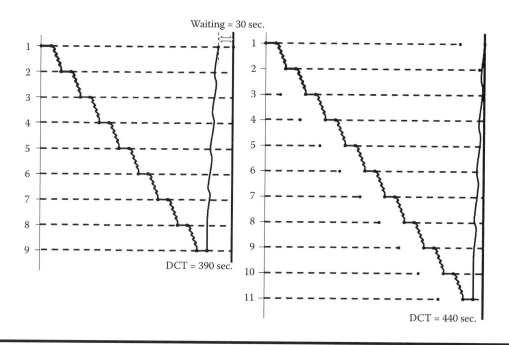

Figure 5.10 Comparison of WCT graphics for varying the number of parallel machines by one.

If one of the machines were removed, the total work would drop from 400 to 360 seconds (9 × 40 seconds). In this latter scenario, the worker would arrive 30 seconds early (390 − 360 seconds) and wait on the first machine to finish, as shown in the left example of Figure 5.10 (adding 30 seconds of wait for a total of 390 seconds—note that the other machines would continue to run, so there would be no other waiting by the worker, since the DCT is now made equal to the total machine time). However, if another machine were added to the original example, it would yield a total work time of 440 seconds (11 × 40 seconds), and each machine would be idle for 50 seconds (440 − 390 seconds), as shown in the right-hand example. What does all this mean? Referring to Figure 5.10, if we look at the productivity per work cycle, we see that we get 9 parts for 390 seconds of worker time, 10 parts for 400 seconds, and 11 parts for 440 seconds. This yields 43.3 seconds per piece per worker (390 seconds/9/1 = 43.3 seconds) for the nine-machine scenario. The other two scenarios are 40 seconds each. However, notice that the machine idle time is much larger in the 11-machine scenario. It is multiplied by a factor of 11, which would yield a total loss of 550 machine seconds (11 × 50 seconds) per work cycle, which interestingly enough is more than DCT. This latter fact is why the productivity per work cycle did not change from the 10- to the 11-machine scenario. Whenever the worker is performing a 2nd-order or 3rd-order task, the effect is the same as machine idle time if the job is not designed to avoid this.

But suppose the machines do require a minor adjustment from time to time, which is a 3rd-order task. If the adjustments require a technical resource such as a technician, engineer, or maintenance person, then we must add their response time (which can be considerable when resources are limited) to the *actual* adjustment time. Therefore many industries will try to train the worker to take care of minor adjustments, simple repairs, expendable tool changes, etc., to eliminate the response time for a specialized technical resource. If we consider the interface levels involved, we see that the standardized work is ideally composed of 1st-order tasks. On the other hand, we may have designed the job with some 2nd-order and 3rd-order work. One issue that arises from this situation is that whenever the job design includes different interface levels, there is potential for disruption in flow, because any task that is not a 1st-order task could compete for 1st-order task time. In the previous example, the multiple machines represented parallel process flows, so one machine adjustment can easily spill over and affect many other process flows.

In our original 10-machine example, consider that the work content is 400 seconds, so that there is only about 10 seconds per machine for the worker to do the periodic 2nd-order and 3rd-order tasks without impacting the 1st-order tasks that advance the product flow. In fact, flow will cease until the 2nd- or 3rd-order tasks are complete and the 1st-order work continues. But what if we could somehow design the job to allow time for 2nd- or 3rd-order tasks for each machine during the work cycle? Unfortunately, it will translate directly to waiting by the worker on every machine that does not require them. This is because wait time will not accumulate, and therefore it is wasted for every cycle that is not disrupted. If we allowed 50 seconds of excess machine capability for each cycle in the job design (by adding an extra machine) "just in case" a minor disruption occurs, as we did in the 11-machine example, it would still only protect the flow for 50 seconds (as we learned in our discussion of buffers) and could prove to be quite expensive, depending on the cost of the machine. The extra machine capability is not going to help the situation, since the worker cannot operate the other machines, a 1st-order task, as they will be performing 2nd- or 3rd-order tasks on the machine in question. Consider that the extra 50 seconds of protection would be used up on the other machines while the worker was dealing with the disruption on the particular machine, since they would continue to run until complete, so an extra machine would only allow 50 seconds of protection total for all 11 machines—not an economically feasible situation.

The point of this discussion is to illustrate the effect of combining 2nd- and 3rd-order tasks with 1st-order tasks in worker job design. Note that the situation changes if the 1st-order tasks are not disrupted (for example, having a different worker perform the 2nd- or 3rd-order tasks). Therefore, the job design must consider the effects of disruptions when mixing interface levels, because it can lead to waste that is included by design. And since there are 11 different product flow paths involved rather than one path in a normal cyclic application, the effects are multiplied by a factor of 11.

It should now be obvious that one way to deal with a situation like the one just described is to try to design the job so that it does not mix interface levels. For example, it might make sense to assign the technical resource work to another worker job design, one that is designed for noncyclic tasks (which we will describe in the next section), such as the 3rd-order work, which often causes longer and more random disruptions. This way, the 2nd-order tasks could be the focus of kaizen activities to reduce their duration while combined with the 1st-order tasks.

Striking a Balance: Man and Machine

Before leaving this section, a short discussion on balancing human work and machine work is in order. It is often a lively debate in many companies concerning whether it makes more sense when designing work to strive for equipment utilization over labor when there is a choice. There are many good points to both sides of the argument. Consider the multiple-machines example just discussed. If the equipment is expensive, it is very tempting to make sure the machines are highly utilized, and the available worker idle time can be used for 2nd- and 3rd-order tasks. However, if the equipment is not real expensive or if there are other resources available for the 2nd- and 3rd-order tasks, it would make more sense to design the job to most effectively utilize the worker's time. The obvious goal should be to seek to achieve the best of both resources. The problem with achieving such a goal is that there are many processes in industry that do not lend themselves well to capacity flexibility when that process is combined with other standard processes. This usually results in one or two processes that have fairly expensive equipment in comparison to the others. Often, the equipment capacity does not match well with the other processes, and the question of machine utilization of the expensive equipment versus utilization of the worker arises. It is very tempting indeed to opt for higher utilization of the high-cost processes, since it seems like we are receiving better value for our money—but this may not actually be the case.

Shigeo Shingo (1989, 72) describes why Toyota looks first at worker utilization, even at the expense of an idle machine, because they feel that, for most processes, the worker costs are more than those of a machine. Basically, he explains that equipment seems more expensive because of the upfront costs such as initial investment and accelerated depreciation. This can lead to a false assumption that the equipment has no value after it is completely depreciated, so that we may feel pressured to use the equipment while it still has "book" value. His reasoning is that the equipment value should be based on the value it has to the company rather than on the financial accounting value. Take, for example, a machine that initially costs $600,000 USD (U.S. dollars) and is fully depreciated in 10 years at a rate of $60,000 per year. Because the equipment depreciates regardless of whether it is the weekend, or only a one-shift operation, this yields a depreciation cost per hour all year long of about $6.85 ($60,000 ÷ 365 days ÷ 24 hours/day = $6.85/hour). This cost also does not account for use of the asset *after* the 10-year period. If it did, the cost per hour would be even lower. So we begin to see that although the financial accounting system may not always indicate this, the equipment continues to have value over the entire time it is in service, and we can feel obligated to run the machine as much as possible during the depreciation time, even though we may not necessarily need the parts at that rate (which could lead to overproduction if we try to obtain the highest utilization of the asset).

Now compare the above figure of $6.85 to a worker who is paid $30 per hour, including fringe benefits. It becomes apparent why Toyota would feel that the worker utilization is more important under most circumstances. Also note that the equipment cost per hour is the same whether it is used or not—it is a fixed cost, whereas workers are only paid for the time that they are scheduled to be working—a variable cost. Another point to note on why Toyota feels that worker utilization is more important than that of a machine is that one of the basic philosophies of the Toyota Production System (TPS) is the respect for people (Dennis 2002, 143). Many people we have known consider that it is disrespectful to them for the company to waste that person's time in order to better utilize a machine, at least without acknowledging the fact and the reasoning behind it.

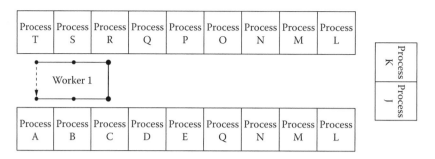

Multiple Workers In Cell:
- Diagrams represent the walk planned for each worker
- Dashed lines indicate return to starting point–expected empty-handed walk
- Heavy lines indicate planned empty-handed walk that is not the end of the cycle
- Normal lines indicate planned walk while transporting part to be exchanged at next location
- Assume that there are handoff points between adjacent workers

Figure 5.11 Basic rules for sketching multiple workers in a work cell.

In some cases, we may have no options to asking the worker to do something that adds no apparent value (waste), especially when walking is involved. Consider that if there are a lot of tasks to be performed, it often requires more than one worker in a concerted effort. In some instances, there may be several workers required. When this occurs, multiple hand-offs are often required as well, which increases the waste in the form of increased SIP total and additional part movement for the SIP. Referring to Figure 5.11, assume that the worker carries a part in one hand and performs an unload operation with the other hand on approach while simultaneously loading the part that he or she was carrying. (Again, for the sake of simplicity, we will remove SIP from our examples when practical.)

Notice that the person's work sequence has the worker traveling across the work cell to the other side. This is often necessary because the total of the task times is much larger than the DCT. So if this is the case, we must add additional workers to ensure that the work is done within the allotted time. When adding more workers to the system, there will be some new issues that arise, such as how the work is divided. It is also very easy for waste to enter into the work because the tasks may not distribute evenly among the workers. Also, we need for the worker to be performing work that adds value for as much of his or her work cycle as possible. Notice in the previous diagram that we have introduced a new notation for crossing to the other side of the cell to a nonadjacent process location (usually empty-handed as well). This is similar to the return to the start of the cycle empty-handed, as that is expected. However, in this case, we have introduced this notation to make visible that the worker is doing so empty-handed *by design* (making the problem visible). We are doing this to indicate a break in the worker flow from the product flow. Most often, this occurs when a handoff point is involved in a sequential product flow such as a work cell. Now let us return to the example.

Suppose that each process requires 5 seconds of worker time. Then for 20 processes, each requiring 5 seconds of worker time, the total work equals 100 seconds. If takt time (TT) is 30 seconds, then there is enough work for $100 \div 30 = 3.3$ workers. But this does not include any walk time. In this example, we will round up from 3.3 to 4 workers. If we assume there is about 11 seconds of walk per each worker, this will yield 44 seconds of total walk ($4 \times 11 = 44$ seconds total). If we add this to the total work, we have 144 seconds, or 36 seconds ($144 \div 4 = 36$ seconds) per worker. Now we divide 144 seconds by 30 seconds, and our theoretical number of workers becomes 4.8, but we will follow the process first with four workers to try and understand the principles involved. Even if we somehow could divide the work and walk evenly between the four workers, this is much larger than the takt time, as shown in Figure 5.12.

The 11-second estimate of walk per worker for four workers was arrived at based on some assumptions rather than actual observation for the purposes of illustration. In our example, we are assuming that each process requires the same amount of worker time, so let us further assume that each machine is the same size and that the distance between adjacent stations is equal (we arbitrarily used 2 seconds for the walk between two adjacent machines). This is to eliminate differences between process steps for this example. One last assumption that we will make is that whenever the worker must change walking directions, some small additional time is added. And this time increases with the abruptness of change based on the number of partial steps and repositioning of their feet and bodies in order to proceed in that direction. There may also be ergonomics considerations.

We will take a few moments to consider how this is relevant shortly, but first continue with how we arrived at the 11 seconds of walk for this example. In the given example, we added another worker after rounding up from the theoretical number of workers based on the ratio of the total work to the takt time as a starting point. To determine which five stations a worker will handle, we will consider those that are closest in terms of walking distance. Because we want to make sure that the system does not contain more

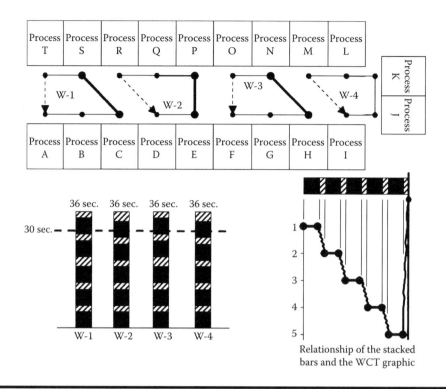

Figure 5.12 Example of job design after rounding up from theoretical number of workers.

product than it needs to function as intended, it is best to have the starting and ending processes handled by the same person. In order to link the rates at which the products enter and exit the cell, we will start with those processes. If we try to follow this same idea of minimizing the walk, the four walk patterns could look like those in Figure 5.12. If we use a base assumption that the straight-line walk time between the stations is 2 seconds, we see that the workers would have to make at least one abrupt change of direction per cycle. (In this case, we are referring to abrupt changes as those that require the worker to turn more than 90° from the previous direction of travel). If we add 1 second for each abrupt change of direction and the small additional distance traveled, then the total walk time per pattern is now 11 seconds ([2 × 5] + 1 = 11).

However, for the sake of simplicity in this example, we based our estimates mostly on the walk distance, such that a walk directly across the cell will be roughly equivalent to the same distance to an adjacent station if we assume that the aisle is narrow enough. The theoretical number of workers was 3.3, which we rounded up to 4. Dividing 20 stations by four workers yields five stations per worker. Looking for the shortest pattern for a worker with five stations resulted in a group of patterns as shown in the layout portion of Figure 5.12. This is how we arrived at the 11-second number for the sake of this example. In actuality, most processes are not the same cycle time, nor are most machines the same size and shape. The reasoning for this series of examples is to take these differences out of our considerations while we look at the work-balancing issues. In real life, the differences in walk distances due to machine size or differences due to process-time mismatches also affect worker job design, but we wanted to simplify the example for discussion purposes.

Returning to our discussion of an abrupt change of direction in a walk pattern, consider the example in Figure 5.13, where we show another option. There are still four workers, each handling five stations, although in different patterns. In this example, if we use the same assumptions for the walk times, the results for each worker might be as shown in the lower portion of Figure 5.13.

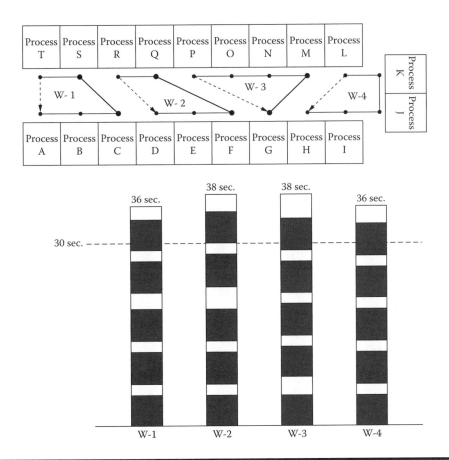

Figure 5.13   Example of impact on walk time when trying to "balance" work time.

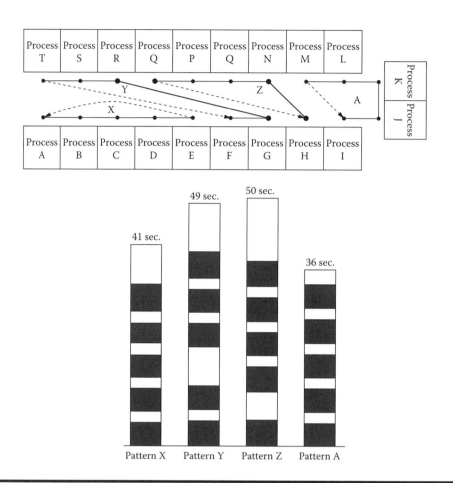

Figure 5.14 Some extreme examples of walk time impact of job design.

The extra distance caused by trying to have the work pattern include stations farther away—often as a result of trying to achieve a "balance" between all the workers in the cell—can become significant, as shown in the next example (refer to Figure 5.14).

Consider the four example walk patterns in this illustration. All four patterns include five stations for the worker. Pattern X is five side-by-side stations all in a row, with a walk back retracing the original path. This results in a walk time of about 16 seconds (2 × [2 + 2 + 2 + 2] = 16 seconds). Pattern Y has very abrupt direction changes and results in about 24 seconds (using an estimate of 9 seconds to make the long walks and the abrupt change of direction). Pattern Z is similar but makes a long walk to reach a single station before returning to its general region (using an estimate of 6 seconds and 13 seconds, respectively, for the two crossover walks), for a total walk time of about 25 seconds. Pattern A is the 11-second walk from the first example of this series. If the 25 seconds of work for the five stations is added to the walk totals, as shown in the lower portion of Figure 5.14, the worker totals for patterns X, Y, Z, and A, respectively, are 41, 49, 50, and 36 seconds for the same amount of work! Trying to balance the workload with work that is too far away can obviously add waste in the form of additional walking, which in turn increases the workload required. Therefore, the best option is usually to group the work in the desired locations and use the imbalance to create the pull for kaizen efforts.

We can now see some general rules starting to appear:

- Minimize walk between adjacent worker locations
  Watch for changes >90° when changing direction
  Make stations as narrow as possible
  Eliminate or minimize gaps between processes
- Avoid walking back over ground just covered when not performing work
- Minimize walk without part if the worker is the means of transport in the cell (Notice that pattern Z has stations along the path allowing work as the worker crosses the cell; this is one of the reasons a cell is laid out in a *U* shape.)

There is another issue to point out here. If the 13 seconds of walk difference between Y and Z was about 30 feet and the cell had to produce 240,000 parts in a year, the total extra walk equals 7.2 million feet or 1,363 miles! And this adds absolutely *no* extra value! Another way to look at this is that 13 extra seconds of time per part for the year is 3.12 million seconds or 866 hours of extra *wasted* worker time! Some of this time is not even carrying a part (note the heavy lines on the example patterns).

However, since we now have confirmed that four workers will not work in the original example, the next step is to add a fifth worker. We design the patterns for five workers following the same process as we did for four. The 20 processes divided among five workers equals four processes each. In this example, the walk time per worker is about 8 seconds, or 40 seconds total. The total work and walk time is then 140 seconds. This works out to a time per worker of 28 seconds (140 ÷ 5 = 28). The new pattern group and resulting graph is shown in Figure 5.15.

The five workers are "balanced," but there is a problem in that we are designing in 2 seconds of idle time per each worker (recall that the takt time is 30 seconds). We could argue that this would provide for some allowance for losses, so that in fact the 28 seconds is actually the desired cycle time

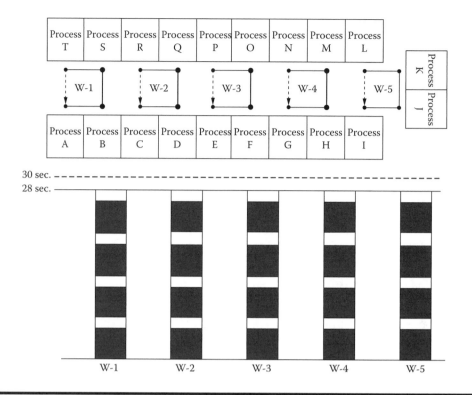

Figure 5.15 Job design for initial example with additional worker.

(DCT) and the resulting ratio of DCT to TT is 93%, or a 7% allowance. This may not be enough to account for a reasonable amount of losses. However, for the sake of simplicity, we will use this DCT for this series of examples. Assuming for the moment that this was sufficient, we would need to eliminate at least 28 seconds of total work and walk to reduce the number of workers back to four; otherwise, we would only add idle time or wait to the system when we work on kaizen efforts. However, consider that if the equipment had not yet been purchased and the system had not yet been designed, we could take this opportunity to apply the rules we have discovered to provide a better starting point for kaizen by eliminating as much waste as possible *before* we install the system (and possibly before we actually spend the money).

Suppose we look at a slightly different example where the change is that there now are 18 processes instead of 20. If the total work is 90 seconds and the takt is still 30 seconds, there is enough work for three workers. From our earlier discussion, we know we need to allow for walk when there are multiple workers involved. However, it is more difficult to determine the walk for each worker until the diagram for each worker is known. How should we proceed? We follow the same basic process. If we use the rule of thumb that the base number of workers is total work divided by takt and then add one more worker (since we only eliminated 2 stations out of 20), we again start with four workers. But we note that unlike the original example, which was evenly divisible by 4, this example is not. Can we obtain an "ideal" solution? Most likely not, as there are seldom any ideal solutions, and when they appear ideal, it is only for a short time if we are using the kaizen process. If we look at the various walk "patterns" we had in the previous examples, we can come up with a reasonable estimate. If we assume that we have four workers, then two workers would have five stations, and the other two workers would have four stations. With the walk times observed previously, this would be about 38 seconds of walk (11 + 11 + 8 + 8 = 38). Now the total work and walk is 128 seconds. If we start out with one of the other rules we discovered—that we should include stations that are within the shortest distance—we might end up with a group of patterns and the accompanying graph, as shown in Figure 5.16.

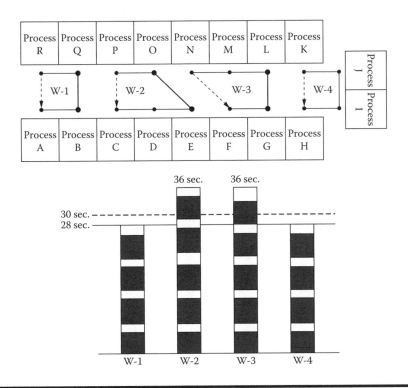

Figure 5.16 Example starting with theoretical number of workers and two fewer workstations.

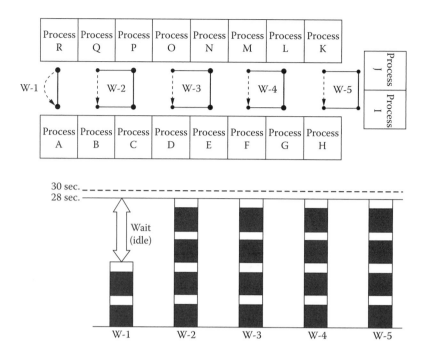

Figure 5.17 Job design for second example with additional worker.

In this example, we see that it is impossible to "balance" the work between the four workers. In addition, two of the workers are 2 seconds under takt, but the other two workers are 6 seconds over takt! This system cannot meet the output requirements. If we proceed as we did in the earlier example and add another worker, the new group of patterns might look similar to those in Figure 5.17.

By following the previous examples closely, we should see that four of the workers would have four stations each, and the fifth worker would only have two. This is an example of how the work might be divided among five workers. Again, we have 90 seconds of work for the 18 stations, and the walk is about 36 seconds, for a total of 126 seconds. Only 2 seconds of extra walk were eliminated when compared to the four-person scenario. The workload chart for this group is shown in the lower portion of Figure 5.17.

Four of the workers are balanced with the DCT established in the earlier examples, but the fifth worker is highly underutilized. If we compare this example with the four-person example, which one is correct? There are two main approaches to this type of problem. The first is that the four-person setup is right and that the overtime required would provide the "tension" necessary to fuel the kaizen efforts. The other is that the five-person setup is right and that the visible waste of the underutilized worker would provide the "tension." Each approach has its advantages and disadvantages. However, a blend of the two is probably the best approach, depending on the amount of kaizen required, the availability of resources, and other factors.

One last point to note before we continue is that, in this series of examples, we considered the machine cycle times, the walk times between the machines, the size of the machines, and some other facts to be equal between processes in order to easily eliminate them from our discussion on worker job design. In reality, machines and workstations come in all shapes and sizes, process times can vary considerably, and walk times depend on a myriad of factors. If we were to add these facts in, the process would be similar, although we would just need to allow for this in our analysis. For example, when trying to develop the number of processes that a worker would be assigned, it should obviously be based heavily on those within the closest proximity. The first and last operation should be performed by the same worker to ensure that the pace of products entering and exiting the system is equal. The total walk time around the system, as determined by a single worker going through each process, can give you the beginning basis for the total walk required, although extra time must be added to compensate for crossing over the cell to reach nonadjacent processes. If the total work time is added to the total walk time and then divided by the takt time (or DCT), a theoretical number of workers can be determined as the starting point. The process then continues as discussed previously.

Some New Rules and Some New Tools

During this last series of examples, we discovered several general rules that may be helpful in job design for cellular-type scenarios. They are:

- Use diagrams to represent the walk planned for each worker
  Dashed lines indicate return to starting point—expected empty-handed walk
  Heavy lines indicate planned walk to a nonadjacent process location
  Normal lines indicate planned walk to an adjacent process location
  Assume that there are fixed handoff points between adjacent workers
- Minimize walk between adjacent worker locations
  Watch for angles <90° when changing direction and allow time for abrupt changes in direction (see Figure 5.18)
  Make stations as narrow as possible

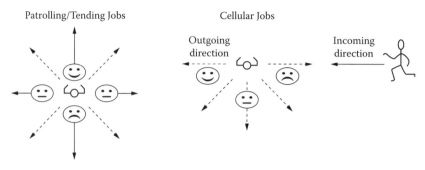

Figure 5.18 Impact of abrupt changes in direction of worker job design.

Eliminate or minimize gaps between process locations

Arrange the processes in a pattern that allows transfer of work but supports smooth product flow

- Avoid walking back over ground just covered when not transporting a part
- Minimize empty-handed walk when the worker is the means of transport in the cell
- The same worker should handle the first and last processes to link the product going into the cell with the product leaving the cell

One final point to make before we continue on with our discussion is that there are some simple tools to compare the workloads of the workers as well as those of the machines. Figure 5.19 is an example of how the last example in the previous series might appear. Notice that we have entered an

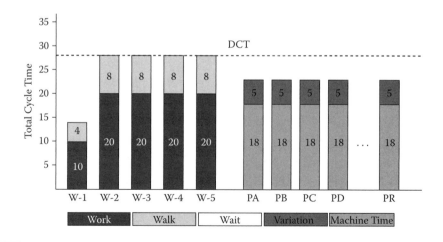

Figure 5.19 Simple graphical tool for depicting worker and machine workload.

arbitrary cycle time for the machines of 18 seconds each. Using this chart, we see that the DCT is actually driven by the workloads of the workers. If we could somehow reduce through the work or walk content—either by redistribution of some of the workload onto W-1 or by kaizen efforts on the workload of W-2 through W-5—there is the possibility of reducing DCT to as low as 23 seconds. We also notice that the machines are underutilized, which is what offers this opportunity. Therefore, a simple visual tool can be a very powerful means of pursuing kaizen efforts. Also note that the legend on the tool defines wait and variation, even though these were not used in our example series for simplicity (see Figure 5.19).

On the subject of workload or work-balance tools, we would like to point out that they should be as simple as possible and still show the important points. The reason for this is that it is very easy for a calculation-based tool to be made more complex than it needs to be. There is a lot of interesting information that can be derived from mathematical analysis of the different aspects of trying to obtain an optimum workload balance. However, it should be noted that the ultimate purpose of such tools is not to achieve work balance in itself, but rather as a tool to support the kaizen attitude. Therefore, it is essential that problems be kept visible and that the concept of ultimate ideal solutions be disregarded in favor of continuous improvement. If this is not the case, it is likely that the pursuit of perfection of the tool itself will become the goal.

As a last example to consider in our discussion of job design, we will look at a surface-mount operation (normally referred to as SMT in the electronics industry). These types of operations often mix 1st-order, 2nd-order, and 3rd-order tasks in the job design. In Figure 5.20, we try to show the workers' tasks and their location relative to the equipment and each other. Recall that in our earlier discussion of long-cycle standardized work, we introduced two other notational conventions for the layout sketch to be used for the SWC.

Basic Layout Sketch Rules:
- Sketch a rough approximation of each workstation and its location in relation to the other workstations.
- Each major work sequence step is shown as a number within a circle.
- The steps are connected by solid arrows.
- The return to the step that begins to repeat the sequence is shown with a dashed arrow.
- Work step, equipment name, etc., can be shown for clarity.

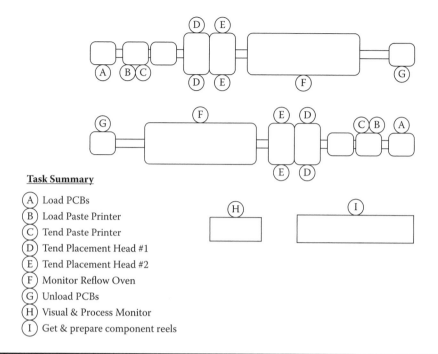

Task Summary

(A) Load PCBs
(B) Load Paste Printer
(C) Tend Paste Printer
(D) Tend Placement Head #1
(E) Tend Placement Head #2
(F) Monitor Reflow Oven
(G) Unload PCBs
(H) Visual & Process Monitor
(I) Get & prepare component reels

Figure 5.20   Layout sketch and task summary for surface mount example.

■ Parallel tasks that do not have to be performed in a particular sequence compared to other steps—but simply need to be performed on demand—are denoted by letters.

■ A combination of sequential and parallel tasks is denoted by using numbers, letters, or a combination of both.

In Figure 5.20, we can see a mixture of all three interface levels. Job tasks A and G are 1st order, even though the printed circuit boards are moved in magazines. Job task B is a 2nd-order task (replenish solder paste), and job task C is a 3rd-order task (process adjustment). Job tasks D and E are also 3rd order. The remaining job tasks are probably all 2nd-order tasks. Notice that whenever the worker is required to do a task that is not A or G, that worker is unavailable for 1st-order tasks. The magazine handlers allow for several minutes of protection for disruption, which allows the worker time to do the 2nd-order tasks (splicing component reels, performing quality checks, etc.) and 3rd-order tasks (machine and process adjustments). In the given example, suppose that there are 10 SMT lines located in close proximity. It might be possible to assign all the 1st-order tasks solely to one worker so that the flow of material did not stop due to the worker being unavailable. In fact, the technical training of this worker would not have to be as

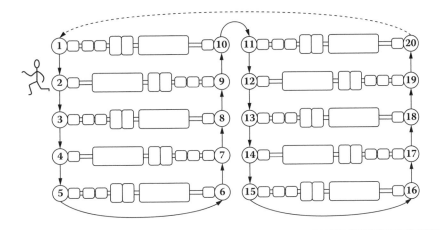

Figure 5.21 **Layout sketch for loading/unloading ten parallel surface mount machines example.**

extensive for the 1st-order tasks as would be required for the 2nd-order and 3rd-order tasks. The layout sketch for this scenario is shown in Figure 5.21.

The work-combination table (WCT) for such a job design would be fairly large. In this scenario, there are 20 separate locations, 10 load positions, and 10 unload positions. Also note that some of the locations are markedly further apart than the others. Another point to note on this WCT is that— due to the nature of the magazine handlers on each end of the machine, providing several magazines for protection from disruptions—no machine times are shown for the sake of simplicity (although the WCT could be used to help determine the proper number of magazines at each machine for this job). The WCT might look something like Figure 5.22.

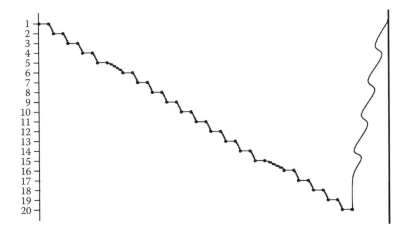

Figure 5.22 **WCT graphic for ten parallel surface mount machines example.**

It is now quite obvious that, in these types of situations, the various inter-face levels can have a great impact on the job design. Many of the conditions do not lend themselves to implementation of standardized work that we have discussed up to this point: cyclic and long-cycle applications. And even if they did, they would be mixed in with tasks that were following no pattern. Unless we can combine several short cyclic tasks, as in the previous example of multiple-machine handling or some other fortunate circumstance, there appears to be no recognizable repeatable sequence to this type of work, or rather it is *noncyclic* in nature.

Chapter 6

Noncyclic Standardized Work

In the previous chapters, we discussed cyclic operations and the basics of standardized work. We also discussed how these same basics can be applied to situations that did not seem to lend themselves readily to short-cycle standardized work methods and where the takt times tended to be fairly long when compared to the work period. We learned that it was still important to do the work steps of each individual task the same way each time they were required, even if the *exact order* of tasks themselves were not repeated in a constant loop. Work sequence also becomes less clear if we continue to compare things to the way they are in short-cycle applications experienced in manufacturing cells where a worker typically completes the same series of tasks over and over for their entire work period. Not all work can be accomplished in such a simple and straightforward way. Therefore, it is necessary to look at things from the perspective that the worker may not complete many repeated cycles within a single work period. It is important not to lose sight of the overall goal by looking at things too closely. This is similar to the old saying that a flea living on an elephant does not really see the entire animal but only the small area where it resides. Sometimes we must look at things from a wider perspective. When we do this, we begin to see that the concepts for the simple short-cycle applications still apply. However, in this type of situation, it is necessary to look at takt time from a different perspective, because there may be multiple product value streams involved. Consider the model shown in Figure 6.1.

This is similar to the cyclic and long-cycle standardized work models, except for the fact that we "set" the takt time to be equal to the work period. In other words, since there may be multiple product value streams involved

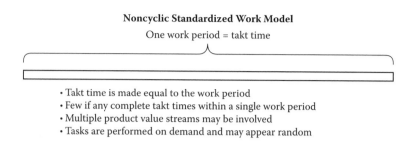

Noncyclic Standardized Work Model

One work period = takt time

• Takt time is made equal to the work period
• Few if any complete takt times within a single work period
• Multiple product value streams may be involved
• Tasks are performed on demand and may appear random

Figure 6.1   Noncyclic standardized work model.

or the worker may be performing tasks that apply to different products, individual product takt time plays less of a role in determining whether we are ahead of or behind where we need to be, since the takt times between products can vary greatly. Actually, in cases like this, the efficient utilization of the worker's time within the work period takes on more of a role, since the advancement of multiple products toward completion along their individual value streams is dependent upon the worker job design being focused on adding as much value as possible. So it is in this context that we look at noncyclic tasks and the application of the principles of good standardized work that we have discussed.

Parallel Work Steps or Tasks on Demand

However, at this point, we see that the essence of the work in applications like the SMT (typically referred to as a surface-mount operation in the electronics industry) example in the last chapter is much different than the previous discussions. The work appears to completely lack any repeatable work sequence unless there is a chance for multiple machine handling. In one sense, this is correct; the work occurs *on demand* from the nature of the product, process, or worker rather than on the customer's schedule. But from another perspective, we could say that the sequence does exist; it just happens to be *random* (or at least it appears that way to us at the moment). The important points for the worker are:

1. Recognize when a task is required to be performed.
2. Recognize what task is required.
3. Recognize what work sequence is required for the task.
4. Recognize when a task is complete and return to point 1.

Workers on Patrol

Situations like this are not unique. One that comes to mind immediately is that of a computer program waiting for input from the user. During normal operation a continuous loop is performed, monitoring many conditions and waiting for a triggering event. When that trigger occurs, a decision as to what subroutine is required is made. Once the subroutine is complete, the program returns to the continual looping, waiting for the next event. An example from the real world that also comes to mind is that of a soldier patrolling a defined area. During the normal patrol, a triggering event occurs, indicating that action is required. At that time, training takes over, and the soldier performs the required steps. Once the event is over, the soldier returns to the previous patrol. If this analogy is applied to our noncyclic work applications, we would observe a situation similar to that shown in Figure 6.2.

Now let us consider how to apply standardized work principles to work that does not seem to be cyclic at all, or is possibly a mixture of short-cycle, long-cycle, and noncyclic tasks. Some industries must produce in batches, and these batches may require processes that take longer than a single work period. Often, workers in such industries will handle or "tend" multiple machines, as we learned in the SMT example in Chapter 5, or even multiple processes of different product flows as necessary during the course of their work period. They may or may not interact with all batches on their particular shift, although some batches may require several interactions, depending on where the batch is in its manufacturing sequence and the interaction time required during any event that occurs. Such conditions are common in some industries, such as integrated-circuit (IC) fabrication.

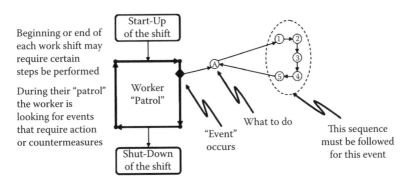

Figure 6.2 Example of worker "patrol" concept.

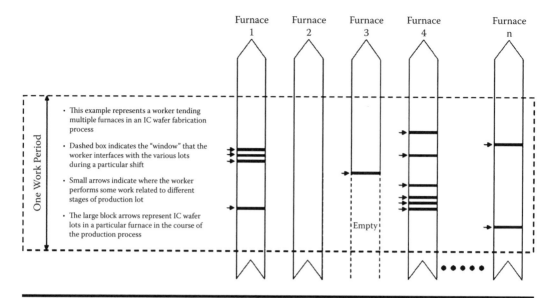

Figure 6.3 Example of worker interaction with parallel furnaces running different profiles.

In the example in Figure 6.3, takt time for each product flow is still defined similar to discrete parts: using the time allotted and the number of lots required within that interval (in some cases, this is much longer than a single shift). In this case, takt would be more similar to a schedule of several batches, since the batches in a furnace are long-cycle in nature, and the worker is tending multiple furnaces simultaneously. However, takt time applies to the product, and since there are multiple batches of the same or even different products being produced simultaneously, takt time seems less appropriate at the moment than good job design for the worker—in other words, the worker is well utilized with the focus on 1st-order tasks (adding value and moving the product closer to completion). This is really the same as we discovered for short-cycle standardized work applications; the worker time is intended to be utilized effectively for the entire takt period. The difference here is that the takt time for our purposes extends to fill the entire work period, so the main objective is to strive for effective worker time utilization and smooth product flow while relentlessly pursuing the elimination of waste. Recall that for long-cycle applications, the work period had exchanged places with takt time as the smaller unit in Figure 6.3. It is often difficult to even consider takt *time* in the conventional sense, since the worker is running multiple machines in parallel, and the products may actually run through the same process at different parameters at different stages of their manufacturing

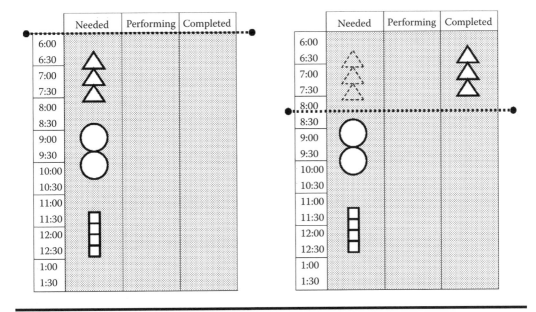

Figure 6.4 Slight variation of earlier visual takt time example with planned batching.

sequence (and possibly creating a very complicated process flow). Thus, in most instances, the concept of takt still applies to each of the products, but it is difficult, if not impossible, to measure in time units that directly relate to the worker's time, so the visual approach seems to be called for in these cases.

Recalling the example from our discussion on what takt is really trying to convey, suppose that what we need is a visual tool for a surgery team. The team has an operating room and equipment scheduled for an 8-hour period from 6:00 a.m. to 2:00 p.m. Referring to Figure 6.4, if the symbols represent the different procedures planned and the blank spaces between them are times allotted for preparation and cleanup (changeover), then the visual approach can indeed tell us what we need to know: Are we ahead of or behind where we need to be? This information is helpful so that, if difficulties are encountered, arrangements can be made for more time, more resources, rescheduling of later procedures, etc. We extend our apologies to the medical profession if we overly simplified this example to make a point.

Merging Takt Time and the Work Period

Returning to our discussion, note that if the job is designed correctly, the worker will be utilized effectively for the entire work period. Although the takt *time* is difficult to quantify, if we utilize a visual scheduling-type approach, the work elements for each step in the sequence are still the same each time they are required. Another twist is that, in some cases, the product may go through the same process more than once and with different requirements. In our IC fabrication example, since batches of wafers may pass through the process multiple times, the worker must be able to perform the required task each time, which requires some additional decisions on the worker's part. Also note that the work content very likely may include some 2nd-order tasks, such as replenishment of materials, process checks, etc., as well as some 3rd-order tasks, such as machine or process adjustments. In this situation, the worker is performing periodic tasks required for very long cycle processes in parallel. Assuming that the job is designed correctly, the focus should be on performing the tasks in order of priority: 1st order, 2nd order, then 3rd order. The role of takt time for this case takes on the form of the work period: Good job design facilitates the proper priority, and the measure of this is the effective utilization of the worker's time during the work period. Consider Figure 6.5. The left bar graph represents the work-element summary of the worker's time for a work period. The right bar graph shows how the work portion may be designed or planned out at the task level. The block at the top represents the opportunity for improvement, which means that a reduction in walk or wait allows more work to be

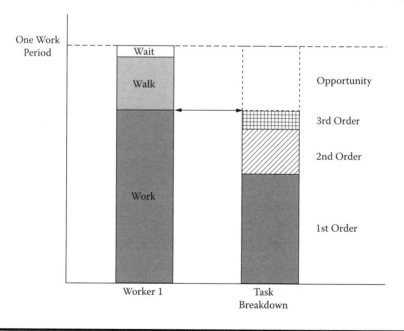

Figure 6.5 Example of task breakdown and improvement opportunity.

accomplished, thereby increasing the effectiveness of the worker's job. The graphical tool does not ask the worker to work harder. It simply helps to make the worker's time more productive, thereby not wasting worker time.

On this last point of making the worker's time more productive, we would like to point out that it is not our intent to overburden the worker to the state of exhaustion, which would be disrespectful to the person. This is another contributor to waste. If the worker is too tired or fatigued, the result will be variation. For example, consider the situation where a worker begins the shift fresh and rested. The total time taken may be much less in the beginning of the shift than near the end of it if the work pace cannot be comfortably maintained. Therefore, it is important to remember that, for good job design, this must be taken into account. The pace that a worker is asked to maintain should be designed so that it *can* be maintained. It is for this reason that we want to mention the importance of ergonomics and human factors in the design of good standardized work. There are many sources where information on these subjects can be found. However, it is essential that we recognize that these should be a consideration when we are trying to design the worker job. It may be necessary to redesign the workplace, the tools, and equipment, or simply rotate workers between different jobs that utilize different muscle groups.

Assisting the Worker: Standardized Work Drives Equipment Needs

Even if the job is designed correctly, how does the worker recognize when something needs to happen? Should the worker constantly be looking for what to do next? From a certain point of view, the answer is yes. The goal should be to make it extremely easy to recognize when something needs to occur. For example, the magazine handlers on the ends of an SMT line can have a stacked-light system that begins to flash when the handler gets near the point of needing worker intervention in order to attract the worker's attention. However, it gets more difficult to "alert" the worker when 2nd-order tasks need to occur. This requires some work in the job design to do this in a simple and economical manner. Sometimes, it depends on whether or not the 2nd-order tasks can be delayed for a short while. Consider a quality check that is required to occur every time 1,000 pieces are run. Does the requirement dictate that the worker verify that the 1,000th piece has been checked before any others are run? If so, the machine could be programmed to shut down with a reminder for the worker. If there is a little room for leeway, perhaps the machine can simply be programmed to alert the worker that a check needs to be performed, and the worker can reset the alert once the check is complete.

The key is to understand the requirements and use the equipment and tools at hand to aid the worker in recognizing when a triggering event is about to occur or has already occurred—without overly complicating the equipment. Perhaps the event is based on time, such as a check that is performed once per hour. An aid for the worker that is based on time, such as a timer or even a simple checklist (or status board) that is set up in time increments, might suffice. In any case, the worker needs to recognize when a triggering event has occurred and quickly discern what task needs to be performed. However, it should also be noted that in non-cyclic and long-cycle applications, the recognition of when a triggering event occurs as well as the recognition of what needs to be done can also be off-loaded to someone else, for example a foreman, a dispatcher, etc. Recall the discussion in Chapter 4 concerning the process flows for long-cycle situations such as the construction industry. Some of the process steps have several smaller substeps that can actually occur in parallel or in no particular order (within some portion of the process flow), thus allowing for some adjustments on the fly at the discretion of the worker, a foreman, etc.

Refer once again to the example from the section on long-cycle standardized work (see Figure 6.6). Note that there are some parallel substeps within the major steps. The rules governing the order of priority and other helpful tips and information for decision making can be incorporated into a task summary sheet (TSS). Now the TSS can be used by the worker, foreman, team leader, etc. This helps ensure that the decisions made are done so in a

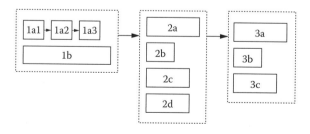

Figure 6.6   Earlier example of long-cycle sequential and parallel work elements.

standard manner, even if they are infrequent or are not repeated in exactly the same order in respect to other tasks each time.

Assisting the Worker: Tools for Complex or Infrequent Tasks

Once the task is identified, the worker needs to know what sequence of steps must be performed. The worker can easily be aided in this last issue by utilizing a tool such as a task summary sheet (TSS). These may be known by many names, but the purpose of the TSS is to summarize the sequence of steps for a specific task that must occur along with other relevant information that the worker may need. Note that this can apply to cyclic tasks as well as noncyclic tasks. An example TSS is shown in Figure 6.7, along with some further explanation of the areas on the tool. The TSS can contain special tricks, tips, or other information that is helpful for the worker to accomplish the task correctly. Whenever there is a TSS for a cyclic task, it is important that the existence of the TSS be prominently displayed on the standardized work chart (SWC) so that the information is readily available to the worker. The information should only deal with aiding the worker in the task and should not define quality nor intrude upon other important technical documentation (refer to Figures 6.7 and 6.8).

Task Summary Sheet This area contains information on the product, process, etc.			
	Major Step	Key Points	Reasons
This area contains pictures, sketches, etc. for assisting the worker	This area lists the major steps that must occur. The small column to the left is used to add symbols or notes	This area lists any key points for a step that should be noted	This area lists the reasons for key points or for the step in general so that the worker is reminded why they are important

Figure 6.7   Example of a simple task summary sheet.

Task Summary Sheet	Toy Truck Assembly Cell		
	Step 3 - Apply decals to both sides of toy truck bed		

		Major Step	Key Points	Reasons
	1	Place truck into fixture with left side up		Saves time
	2	Get left-side decal from dispenser		
	3	Apply left-side decal	Align wide end first	Decal is easier to align from that end
	4	Remove truck and place into fixture with right-side up	Take care to not cause damage to left-side decal	
	5	Get right-side decal from dispenser		
	6	Apply right-side decal	Align wide end first	Decal is easier to align from that end

Figure 6.8 Example task summary sheet for toy truck decal application process step.

Although the TSS is an important tool in the development of cyclic standardized work, it is critical for noncyclic standardized work. There may be some tasks that do not occur very often, and the TSS can help ensure that the task is performed in the same way each time it is required by serving as a quick reference for the worker. It can be posted right at the location where the task is to be performed to serve as an aid to the worker when necessary. The task should be broken down into the major steps. There may be some points or issues that need special mention or instructional diagrams. This can include symbols such as the diamond or cross used on the SWC, or it may be in the form of pictures or illustrations in the left-hand section of the form. The key-points section is used to describe any items of particular concern that the worker needs to be aware of for that step. The reasons section is used to help the worker to understand *why* something is done a particular way when necessary. This can be very helpful when the reason for doing something in a certain manner is not clear or seems arbitrary to a casual observer. It should also be noted here that the TSS is very similar to a Job Instruction breakdown sheet used by the Training Within Industry (TWI) initiative.

A TSS is developed for each task for a noncyclic standardized work application. It is important to not only offer additional information or serve as a reminder for tasks that do not occur very often, but also to capture *all* the

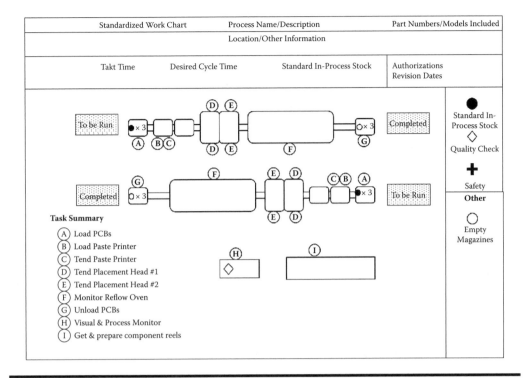

Figure 6.9 SWC for earlier surface mount layout sketch example.

tasks that the worker is expected to perform. This is necessary in order to make sure that all the tasks are accounted for, since noncyclic jobs can easily "pick up" additional tasks that were not intended in the original job design and adversely affect cost and add unproductive tasks. Therefore, it is also important to list all the tasks somewhere on the SWC for the job while still keeping it simple and on a single page. Notice that the list for our SMT example is also shown on the SWC of Figure 6.9 in the lower left-hand corner.

Notice that an attempt has been made to show standard in-process stock (SIP) at the load and unload handlers on the ends of the SMT lines (SMT refers to a surface-mount operation in the electronics industry) in this SWC. The symbol that we defined earlier for use as a handoff point where a part was not normally supposed to be has been redefined to indicate where empty magazines (used to contain completed product) should be for the planned start-up of the job. Also note that there are extra characters (×3) with the circles to indicate that there should be three magazines of printed circuit boards (PCBs) as well as three empty magazines on the unload end. The workstation for step H has a diamond symbol to denote a quality check. There have also been two boxes added to the layout sketch to indicate the location of carts where the PCB magazines are stored at the beginning and end of the SMT lines.

It is important that the SWC for a noncyclic standardized work job show the entire "patrol" area as well as all the tasks that the worker is expected to complete when required. Similar to the work-combination table (WCT), the idea is that problems should be made visible (and sometimes audible if the worker is out of sight of the visual alert). If the area of coverage is too large or there are too many tasks to show on a single page of an SWC, or the tasks are spread too far apart, then this might be an indicator that we are asking the worker to remember too many things, cover too large of an area, or walk excessively. If this is the case, it may be necessary to try to combine this job with another and try to redistribute the tasks in a better fashion. Recall that the larger the distance between geographic locations of the tasks, the more walking is required, which we know is waste. So when designing the worker job for noncyclic applications, it is still important to consider walk times, because although the walk may be infrequent, it still can have quite an effect upon response times as well as competing with 1st-order tasks, and this will impact the decoupling effects of any buffers designed into the job.

Earlier, we suggested ways to assist the worker in recognizing when a task needed to be performed for the magazine handlers and the quality checks. But there are several other points on the task summary list that may need to coordinate with the equipment or visual aids for the worker. Consider that some steps are going to be repeated many times, while others may only occur with very low frequency. The latter will require more thought, as the frequency of performing the task will have a great impact upon the level of assistance that is needed in the TSS. The SWC is meant to be a one-page summary tool, and this is also the intent of the TSS. If it gets much larger than a page, it may mean that too much information is being conveyed, the task is too complex, etc. It is not meant to be a replacement for technical documents but rather as a summary of the steps, in other words, the *method* required for a specific task. The important point about a TSS is that it contains just enough detail to serve as an aid for the task and nothing more.

One other function that the TSS can be used for is the sharing of problems and issues experienced in the past. Over time, there may be quality issues, machine problems, accidents, etc., and it is important to ensure that the lessons learned are not forgotten, even when people who have this knowledge first-hand are not available. If these are briefly documented on the back of the TSS, this can serve as a reminder of the issues that have occurred in the past for infrequent workers as well as new workers when they are being trained. If the information is at least summarized here, it can be a powerful countermeasure to prevent the mistakes and problems of the past from recurring in the future.

Applying Standardized Work to Transactional Processes

The last thing we would like to cover regarding noncyclic standardized work is the application to transactional-type situations. These could be engineering processes and procedures, the processing of a loan application, transactions in a financial department, and so on. The actual form that the work takes is not as important as the ability to visualize the steps in the process. Sometimes, before the work can be defined properly, it must be sketched out, similar to what we might see for a value-stream map. The important thing to remember is that if the work can be described properly, it can be standardized. Although, recall from our earlier discussions, the situation may first require stabilization in order to properly observe and document the work.

In the section on observation, one of the examples concerned the process involved with a purchase order (PO). The initial process was described as consisting of three steps:

1. Originator writes a purchase order.
2. Finance approves the purchase order.
3. Data Entry enters the order into the system.

However, in our initial observations we noticed that there was at least one step omitted, going to the manager to get the okay that a PO could be written. This caused us to question the real purpose of the second step, having Finance approve the PO. This in turn caused us to formulate the possibility that Finance was simply making sure that funds were available to pay for the items on the PO. Our initial observations were reflected on the illustration that followed the example. It is repeated here for convenience as Figure 6.10.

However, once we followed the process to the third step—getting the PO entered into the computer system—we formulated a question that pointed us back to the first step. The question was: How do we know that the items had not already been ordered by someone else and had simply not been received yet? After all, the items will take some time to be received after the PO is entered into the computer system. For example, if these items take a few weeks before they are actually received and in place, what is to prevent

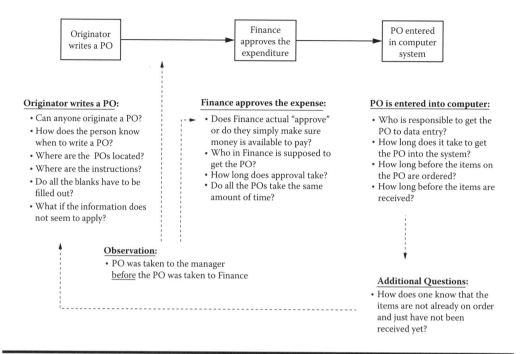

Figure 6.10   Earlier purchase order process observations example.

someone else from ordering them again in the meantime? This is not unlike overproduction. Remember that our observations also generated a question on who had responsibility for producing a PO. If several people could write a PO, then it is entirely possible that items could be ordered multiple times by mistake if there was not a method in place to ensure that this did not occur.

So what does this have to do with our previous discussion of noncyclic standardized work? Simply that the original process, although functional, is not really stable, since it can be performed differently by different people or even by the same person. Processing a purchase order may not actually be the core function of our business, but does this mean that standardized work is not important? That is the first big question. However, the application of Lean works best when it is utilized as a philosophy throughout the entire organization and all functions. Therefore, the waste should be eliminated wherever it is found. However, the prioritization of the waste is where management has the responsibility to make decisions when resources are scarce. But if the kaizen attitude is present and supported, everyone will constantly be looking for ways to improve their job and help the organization at the same time.

So we see that the present condition has quite a bit of variability, because the situation is not very stable. But how do we stabilize the process? If the process is not stable, efforts to improve the process may not have the intended effect. The first step is to go back to the basics of standardized work. Recall the three required components for standardized work:

1. Takt time
2. Work sequence
3. Standard in-process stock

In this situation, since the purpose of the small business was not to simply process purchase orders, then takt time takes on a less critical role than in a cyclic standardized work application. However, the goal should be the timely processing of the PO without an undue amount of waste or delay—possibly a good candidate for using the WCT as well. Therefore, we can assume that the goal is the minimum practical time, because the people involved will have more important tasks where their time is needed, so there is also tension in the process for kaizen.

We will look at the work sequence shortly. The third required component is standard in-process stock. This last one is fairly easy. The requirement is that the PO be processed completely in a timely manner so we would expect *no* standard in-process stock for the system to function correctly. On the other hand, we might consider that the actual purchase order forms need to be present at a specific location when needed (even though one could also argue that the PO forms were actually a raw material).

Returning to the work sequence, we see from our initial observations that the actual work steps are a little different than originally explained. For example, we see that the second step might be better described as two separate steps—getting the manager's approval to make a purchase and then getting the Finance department to verify that money is available to make the purchase. Now the process looks like the diagram in Figure 6.11.

However, this does not make the process more stable, only a little more accurate in respect to what actually happens in the workplace. In the first step, it simply states that the originator writes a PO. It does not specify who the originator should be, what triggers that person to write a PO, where the

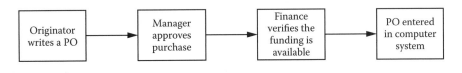

Figure 6.11   Diagram of purchase order process steps updated from observations.

PO forms are located, or how to fill out the PO forms. For our process to be stable, we need to make sure that the individual steps are fairly stable as well. This means eliminating or reducing the causes of variation, and in this case, the elimination of errors (ordering something that someone else already has on order). To do this, it will be necessary to define more details of the *method* to be used.

One way this could be done is to have a single person originate all of the purchase orders. Everyone in the office could update a list of items that need to be purchased that was kept at the originator's location. However, this may not be practical, and it certainly is not flexible. Probably the most flexible way is to allow anyone to originate a PO. However, in order to make this practice more stable, it is probably a good idea to have the purchase orders all originated in the same place, similar to what we were considering in the first scenario with a single originator. This would make it easier to have the PO forms located at a single location rather than have people searching the office for individual "stashes" of forms when needed. Another idea is that a central PO origination site could also contain a list of items that are on order so that there is less chance of the items being ordered again by accident. The items could be removed from the list when they are received, which would be part of the standardized work for the person delivering the purchased items.

Now that this step is a little more stable, it is time to create a Task Summary Sheet (TSS). The TSS can then contain the major steps, key points, reasons for doing things a specific way, as well as information on how to fill in portions of the PO form that do not seem to apply to the items being ordered. The TSS can be posted at the location where the PO forms are to be filled out to serve as a reminder to people who may not fill out a PO very often. And finally, the TSS will tell the originator where to take the PO next—in this example, to the manager for approval. The updated diagram now might resemble Figure 6.12.

The process for originating a purchase order is now a bit more stable and is a candidate for organized kaizen efforts. There certainly were opportunities for improvements before our discussion, but the process was not very stable,

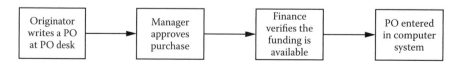

Figure 6.12 Diagram of improved purchase order process.

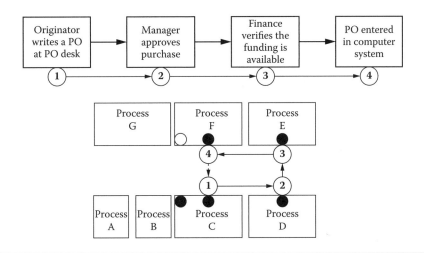

Figure 6.13 Similarities between transactional process and manufacturing process diagrams.

and the efforts may or may not have actually improved the process. The important point to realize here is that the concept and principles of standardized work were easily applied to this transactional process. It was simply necessary to create a diagram of the steps, just as we learned to do for the layout sketch for a cyclic application. The only difference here is that the purchase order process does not continue to repeat in a timely fashion like the cyclic examples, and therefore the layout sketch may look more like a process flow diagram when compared to an earlier example, as shown in Figure 6.13.

This simple example of the application of standardized work to an office environment was meant to show that the concept of short-cycle standardized work, although developed in a manufacturing scenario, can be applied elsewhere with relative ease. The key is trying to understand the thinking behind the principles of standardized work so that they can be applied to your business. The main point to recognize is that in order to properly adapt the principles to nonmanufacturing areas, it is essential that you have a firm understanding of the basics and how they are used in the short-cycle application. This will give you a solid foundation to build upon as you adapt and implement standardized work to the needs of your business or industry.

In Summary

Now we see that standardized work principles can be applied even to transactional processes in an office environment. Although the previous example was very simplistic, it should suffice to show that the principles and concepts of standardized work are not limited to short-cycle manufacturing applications. We have learned that to apply standardized work principles, it is necessary to have a firm understanding of the preconditions:

1. The work must be human capable.
2. There must be a repeatable work sequence.
3. The equipment/tools/workplace must be highly reliable.
4. The materials must be high quality.

It is also important to have a firm understanding of the required components for standardized work. Therefore, it is critical to first understand and have a solid grasp of the concepts of standardized work as applied to short-cycle applications. This is the basis for applying the concepts to other applications, such as long cycle, noncyclic, and transactional processes as well as other nonmanufacturing applications. The required components are simple, although our understanding of them at first glance needed to expand in scope in order to apply them in broader applications. The three required components are:

1. Takt time
2. Work sequence
3. Standard in-process stock

We now understand that takt time is more than a simple mathematical representation of the customer requirements for a predefined time period. It is a *concept* that is meant to help us understand where we are currently in respect to where we should be in order to meet the customer needs according to our plans. If we know whether we are ahead of or behind where we should be, we can institute countermeasures to rectify the situation. We have also learned that the concept of takt is very similar in many respects to a schedule, so when takt time cannot be easily expressed in simple numbers, it is possible to utilize other means to help us determine whether we are ahead or behind. It is best to use a method that is as simple as possible to do this; otherwise, the tools and methods that we use to determine this can become too cumbersome and may actually become a drain on our resources. A simple graphic representation, such as the takt board example, will often work sufficiently for our purposes. Recall the surgical-procedure example presented earlier in this chapter (see Figure 6.4). However, even in this example, in order to make the visual representation tool meaningful, it was necessary to size the symbols to scale with the weekly graph. This in itself is not overly complicated, but it does require a fair amount of work. If too much effort is put into developing a computer-based tool, the tool itself can become the focus of our efforts and limited resources. However, if the tools are developed by hand, a solid understanding of their purpose is much more likely to be developed than by learning to input variables into a software program.

We also learned that the work sequence is the most critical of the required components of standardized work. This is why a graphic representation of the process is necessary. It not only helps us show the real problems in the workplace, as we learned in the importance of geographic locations for a work combination table for a cyclic application; it also helps us more easily observe the process for improvements, especially when the process is not always clear and straightforward. This also extends to the concept of good job design. If the interface levels of the tasks are arranged and distributed properly, it is possible to mix tasks and even product value streams among the worker jobs. This can help better utilize both the worker's time and the equipment and tools.

And lastly, we learned that in order to perform the process the same each time, standard in-process stock (SIP) is a required component. If the job is designed with stock (or the absence of stock) at specific locations, then it cannot be repeated as designed if the SIP is not where it should (or should not) be. The need for SIP can be as simple as a one-piece handoff point in a manufacturing cell, or it can be as large as a 2-hour buffer protecting an internal constraint from frequent disruptions from the preceding processes.

In the course of our discussions, we have introduced several standardized work tools. These include the standardized work chart (SWC), the work-combination table (WCT), and the task summary sheet (TSS). There are more tools out there that are used by others for the purposes of standardized work. However, there are many variations and overlapping functions. Therefore, we have tried to introduce you to the basic tools that we feel are necessary to help you decide if you need additional tools for your business. We have tried to introduce you to some of these other tools by way of some of the simple graphic illustrations we have used. The purpose of these illustrations was to show that a simple visual graphic can often help bring a problem into focus, so that the improvement efforts do not get bogged down in the development and maintenance of complicated computer-based graphical tools. It has been our experience that it is too easy for the focus to switch to the perfection of the tools rather than the relentless pursuit of the elimination of waste. If you cannot do things without formal tools using only pencil and paper, then you cannot do them properly with the formal tools. Besides, using pencil and paper can greatly enhance the observation process and thus our ability to generate more questions to help drive the kaizen attitude.

It is our hope that looking at standardized work from this different point of view will help you understand how these principles and concepts can be applied to any business. Standardized work is not simply a way of documenting the steps of a process; it is a method and an approach that will allow you to not only stabilize your work process, but establish a firm foundation for kaizen in your efforts to improve and become more competitive.

Appendix A: Philosophy for Auditing Standardized Work

This section is intended to provide more information on the philosophy that we recommend when auditing standardized work. It is important to note that the purpose of the standardized work audit is not to enforce the standardized work but, rather, to serve as a friendly reminder that it is our goal to transition to an environment where everyone follows the prescribed method while continually searching for ways to improve. In order to help change our thinking about and adherence to standardized work, the two main goals are to remind the workers of the importance of maintaining the consistency of the currently best known method and to ensure that management is not only aware of this importance, but supports and is actively engaged with the workers in evaluating the current method in order to promote the kaizen attitude. Therefore, the philosophy is simple:

- Serve as a reminder that everyone should adhere to the currently best-known method
- Provide a quick method of checking by making abnormalities very visible for all to see
- Should only be considered a temporary measure until the culture change is complete
- Should be looked at as a helpful process rather than a "policing" action

When trying to audit standardized work, making abnormal conditions visible may take some effort, depending on the type of standardized work being audited. Other workplace tools, tips, tricks, and philosophies from 5S (see Figure 2.3) could help greatly in the visualization of the various methods in the standardized work.

Cyclic Standardized Work

As we know, there are three required components to standardized work: work sequence, takt time (or, in the case of an audit, standard time), and standard in-process (SIP) stock. Therefore, this type of standardized work is very simple and straightforward to audit. The worker cycle time should be roughly equal to or slightly less than the standard or desired cycle time; the work-step sequence should match the standardized work chart (SWC); and the standard in-process stock should correspond to the SWC as well. The first step is to determine what the overall worker cycle-time target range should be (the rate the worker should be producing parts). Although takt time is one of the required components, it represents the customer-requirements target pace and is used as the ultimate standard. In many instances, the actual worker cycle time must run at a slightly faster rate in order to allow for problems, as was discussed previously. However, if the worker cycle time is much faster than takt time, the result can be overproduction. There are generally two accepted measures for this type of standardized work. One is takt time (TT), and the other is desired cycle time (DCT). The SWC will use one of these for the target rate standard. However, if the standardized work is just being developed, the correct target can be determined easily.

DCT is most often used when there is some small allowance for problems being considered or when something in the system limits or otherwise sets the pace for the worker cycle. An example of the latter is where a cell has more than one worker. Suppose that worker 1 supplies parts to worker 2. If worker 1 can only produce parts at a maximum rate of 25 seconds due to the equipment, then worker 2 is limited to 25 seconds, even if 2's cycle is much shorter than 25 seconds. Therefore, the correct standard to use as a measure for both workers in this instance would be 25 seconds, with particular attention paid to worker 2 if the work-combination table (WCT) shows that there should be a wait time because of the limiting factor of worker 1.

DCT is a term used to indicate the rate at which a worker cycle is designed to run naturally, regardless of takt time. The more you have experienced working with standardized work, you will start to see how the "nature" of the work will lead you to a recognizable cycle-time range. An example of this would be a cell where the job takes 20 seconds, even though takt time is 60 seconds. This occurs quite often when a cell runs mixed models, when the equipment increment of capacity is too large, or in the ramp-up/ramp-down of a product life cycle.

Once the worker target standard rate is known, a reasonable range should be determined. Although one of the purposes of standardized work is to reduce variation in the worker cycle, some variation will still occur (this is why we must strive for continual kaizen). A target range should be reasonable but not too wide. Typically, a second or two of variation might be acceptable, but this depends on the length of the worker cycle and the content of the work. After an acceptable target range is determined, the next step is to observe a worker cycle and compare it to the target range. If it is within acceptable limits, then there is no problem with the rate, and the audit should continue to the work sequence. If the observed cycle time (OCT) is outside the accepted range, determine if the cycle was typical or, if something occurred that was not normal and would warrant disregarding that cycle, observe another worker cycle time. Otherwise, there is a problem, and additional observations are needed. This should be noted, and then the audit should continue to the work sequence.

In the observed work sequence, the steps should match exactly with the diagram on the SWC. If they match, there is no problem with the work sequence, and the audit should continue to the standard in-process stock. If the work sequence does not match, this should be noted, and the audit should continue to the standard in-process stock. A few additional comments concerning the work sequence on the SWC:

1. If the worker seems to be struggling physically or mentally, record notes to this effect and review immediately with the worker and the supervisor.
2. If quality items such as visual inspection or gaging are to be a part of the standardized work, then confirmation that these tasks have been completed is critical. If confirmation cannot be verified, then review immediately with the worker and the supervisor.

After auditing the work sequence, the standard in-process stock should be observed. The SWC must indicate the actual standard in-process stock required to perform the standardized work as designed. For example, if a part is expected to be in a station when a worker approaches to begin the work task, this would be indicated by the solid circle symbol for that station on the SWC. Sometimes it is necessary for a machine or station to require several parts to be present to ensure the smooth operation of the process. An example of this is a curing oven that uses a conveyor belt to transport parts through the temperature chamber in a continuous manner. For proper operation, the oven must be full and parts coming out at the proper rate. In cases where several parts are required, the notation should indicate how many parts are required for smooth flow. In the oven example, this could be hundreds of parts. Proper operation may not be possible if the required number of parts is not present. Therefore, it is important to verify that the proper standard in-process stock is in place when auditing standardized work.

Noncyclic and Long-Cycle Standardized Work

It is much more difficult to audit noncyclic standardized work. Some examples of noncyclic standardized work include random frequency events such as machine tending of surface-mount equipment, work that is repeatable but occurs at a multiple of takt time (e.g., every 10,000 pieces), a prescribed number of parts or time period, or even to takt times that are longer than a single worker's work period (long cycle).

Standardized work in noncyclic applications still has the same three required components: takt or standard time, work sequence, and standard in-process stock. Although it is sometimes difficult to perceive these components, they can be the key to properly understanding and auditing this type of standardized work. For example, in a machine-tending job, takt time may not seem to apply in the conventional sense because it may be set equal to the work period. Because the worker is not actually working on a single part, the concept of takt time does not present itself as clearly as it would for a worker in a cyclic job.

Now that we have a better understanding of the nature of standardized work as applied to noncyclic and long-cycle jobs, we can begin to visualize how we can audit the standardized work. Because most jobs that fall into these categories tend to be long in nature, it normally proves impractical to try to audit the entire job. Therefore, the most appropriate method is usually to try and audit portions of the job. The first step would be to locate the SWC and any task summary sheets (TSS) associated with the job. Next, the SWC would be used to locate where in the area and work sequence the person is located and what task is being performed. Once this is done, the auditor can decide whether it is worthwhile to do the audit on the remainder of this particular task or watch for the end of this task and audit the next task that presents itself. If the worker is doing a task that is not reflected on the SWC, this should be noted, and observation should continue until the next task is identified so that it can be audited with the assistance of the appropriate TSS.

We must not forget the importance of standard in-process stock while we try to audit noncyclic and long-cycle standardized work. There may or may not be SIP required at the location where the particular task occurs. However, there may be other materials such as tools, gages, etc., that are required to perform this task, similar to what you may find in cyclic standardized work. We must realize that standard in-process stock does not always refer to productive material in every case. If the worker has to leave the work area because something is missing, this causes a departure from the standardized work. Therefore, it is very important to understand the interface the worker has with the tasks assigned to the job. In some instances, they are directly impacting the flow of the product and moving the parts closer to completion. In others, the worker may be doing tasks that are necessary to support the flow. We have discovered that the level of interfacing is very important to the flow of the product when designing the worker job, and the audit may often have to consider the impact that one task has on the other. Therefore, we can see that even tools, gages, or other supporting materials can be as equally important as products and may actually be considered standard in-process stock.

In summary, the audit philosophy is focused on acting as assistance to the workers and management rather than as some sort of policing action. The standardized work chart (SWC), the work-combination tables (WCT), and the task summary sheets (TSS) should also be a part of the audit process. In a kaizen environment, changes occur quite often. If the SWC, WCT, or TSS is relatively old, this can be an indication that the tools are not up to date. When this happens, this is often an indication that we have a deeper problem and that the audit process itself (which should have already highlighted this issue) may also have some problems. It can be very easy to revert back to an environment where we wait for someone to "point out" that we are not following the prescribed method. We certainly hope that the person to point this fact out is not our customer! The purpose of the audit is to try to prevent this from happening by serving as a friendly reminder: We are all trying to focus on stabilizing the work process to facilitate improvement efforts.

Appendix B: Documenting Common Issues and Problems with a WCT

Now that we have a better understanding of the standardized work chart (SWC) and the work-combination table (WCT), we will try to discuss how to document some of the more common problems that occur when trying to develop a work-combination table. Most problems can be shown by using a combination of the techniques described in the following examples. Sometimes there will be several issues or problems present at the same time for work sequences that already exist. There may be the temptation to start making changes such as rearranging equipment, rebalancing the work loads, and other improvements right away. However, the kaizen attitude tells us that we need to first stabilize the situation before we start trying to make improvements. There are many reasons for this, as we learn from Lean. But it is important to establish a stable starting point in order to gage our kaizen progress. This forms the basis for further improvements. If we want to make many small changes in a continuous fashion, then it is not only unnecessary, but it is also undesirable to make a lot of changes all at once. Consider that if a lot of changes are introduced at the same time, the effects may not be as predicted. But that does not necessarily imply that all the changes were not helpful. Sometimes there are just too many variables that are in play at one time, and changing too many things at once may have effects that are actually detrimental. By changing only one small thing at a time, it does not take a lot of time to determine if the results are as expected. And finally, it is also very simple to simulate the proposed change so that the expected results can be measured without the time and expense of actually making the change permanent. We will explain later on how using a workplace mock-up can be a helpful tool in evaluating change.

One last issue before we consider the examples is that of the importance of using paper and pencil for the documentation of standardized work. The process we have discussed always begins with a layout sketch. We tried to emphasize that very high accuracy and drawing to perfect scale are often less important than just getting the basic ratio and proportion of the layout sketched. Although you may find situations where you believe that accuracy and scale is important, it has been our experience that this is usually not required for the purposes of standardized work. Therefore, we must ensure that other documents or requirements do not creep into our standardized work tools. If this happens, not only can it make the tools overly complicated, but it can also divert our focus from the main purpose of the standardized work—stabilizing in order to reduce variation so that smaller and smaller changes can be implemented on a continual basis. The smaller the change, the less is the risk and the faster it can be evaluated.

However, we do not mean to imply that you simply try any change just because it is small. On the contrary, we highly recommend thorough investigation of any kaizen before it is tried. The main reason for this is that when a change is being evaluated, you should already have a good understanding of what your expected results will be. The more complicated the change, the harder this will be. But fortunately, the reverse is also true, and the smaller the change, the less complicated is the analysis required, and therefore the results obtained should be more definite. You should always make changes based on solid analysis rather than making a change because it "seemed to be better." About the only time when the latter would be appropriate is when there is a high degree of instability in the output of the task and the change is being made to try and stabilize the situation. If this condition exists, it is extremely important to be very aware of the safety and quality impacts of the change.

Getting back to the importance of using a pencil and paper for developing the documentation for our standardized work, it is essential that we go back to our discussion of the layout sketch. Recall that one of the purposes of the sketch was to establish the geographical relationship of the work locations. There can be times when the worker must return to a location or move in different directions multiple times before the cycle repeats. An example of this is shown in Figure B.1.

As we can see from this example, the worker must complete several cycles around the cell before the work begins to repeat. This often happens

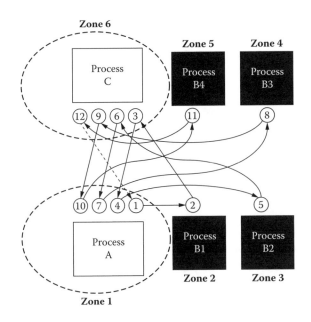

Figure B.1 **Defining geographical "zones" for a WCT with multiple passes before cycle repeats.**

when a process is many times longer than others in the system. In this case, it took four machines to try and balance that process time with the other processes in the system. This is not an uncommon situation. Often a product must be produced within a predetermined target time (or standard). If this is the case, then the total time to produce the product must take into consideration the amount of human work time as well as any machine work time. Also notice that in this layout sketch the worker must return to some locations multiple times in the overall cycle. We have labeled these as geographical zones. The main reason for this will be made clear shortly, when we discuss the WCT impact. As we discovered previously, if the human work time required is greater than the standard time, additional workers must be added. The same is true for machine work time. If the total human work time and machine work time at a particular work location exceeds the standard time, additional machines must be added and operated in a parallel fashion. In some industries, this difference can be considerable, leading to situations similar to Figure B.1. This in turn can cause problems when trying to develop the standardized work tools.

The first problem that we see is with the SWC. As Figure B.1 shows, the layout sketch is very complicated when the actual worker walk path is shown. We have learned that the SWC should be a simple one-page

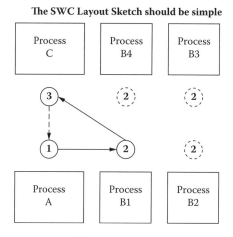

Figure B.2   Simplifying a SWC with parallel machines.

summary of the worker job. If we draw the sketch like Figure B.1, it can be very confusing. For the sake of simplicity and clarity, the layout sketch for the SWC can be simplified greatly by drawing the sketch as shown in Figure B.2. This simpler sketch can still be used for the purposes of the SWC, and the multiple passes around the cell can be easily explained to a new worker being trained for that job.

The next problem we encounter is with the WCT. The simplified layout sketch of Figure B.2 does not show all of the problems as well as Figure B.1. We begin to see now why it is necessary to always start with a layout sketch when developing the WCT. Also, as you will notice from the complexity of Figure B.1, it would be quicker and easier to use a pencil and paper to draw the sketch by hand. A computer drawing is not only sometimes a little difficult, it is not nearly as easy to change the layout while right at the location on the factory floor as it is with a pencil and paper. However, now that we know that we need to start first with the true representation of the layout sketch, we start to get an idea of the additional problems this will cause when developing the WCT. This is one of the main reasons for using the zone labels. Thus, the zone labels will show the actual relationship and impact of the workplace layout on the work sequence. It is necessary to establish how many work steps occur at a particular geographic location so that they can be combined on the WCT, as shown in Figure B.3.

This presents a new problem for the WCT. The actual worker job as designed requires four passes around the cell before all the "B" process machines have been used and the cycle begins to repeat. This affects our

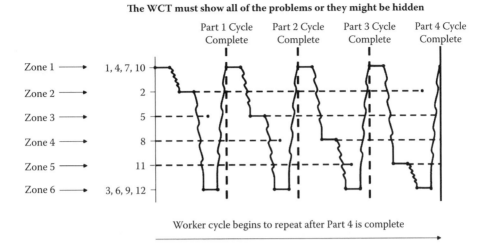

Figure B.3   WCT graphic for example with parallel machines.

concept of takt time as well. The averaging effect of the four "B" process machines in parallel helps to produce one completed part each pass around the cell, but the actual work cycle does not repeat until four parts have been completed. In one respect, this is just a staggered version of operating all four machines in a parallel but simultaneous mode. In this latter mode, four parts would be available at the actual process time, but this would disrupt the other processes in the cell. The number of parallel machines was determined by the standard time and the total human and machine work requirements. By staggering the sequence in which they are used, a common walk pattern can be simulated. Keep in mind for this example that we are assuming that all four machines have the same cycle time. In reality, they could be different. But getting back to our discussion, the only problem for the worker is that it becomes more difficult to know which machine to go to for each pass as the number of parallel machines increases. This is a different issue and may require some equipment design changes to assist the worker in knowing which machine to service next.

However, although we are able to sketch the WCT with a pencil and paper quite easily, there are some problems in trying to use a preprinted paper WCT form. In Figure B.4 we show one method for using a preprinted form, although we must modify the form a bit for our purposes. It is critical that the "zones" be grouped together or the WCT can actually indicate problems that do not exist. Notice that all the zone 1 and zone 6 steps are shown as a single location on the graph portion of the WCT. If they were not, it would appear that they were different geographical locations. The reason that each intermediate step has its own table entry is that there could conceivably be different times for the different steps based on physical layout, etc. Also notice that we have modified the preprinted form by sketching the dotted line boxes on the far left side of the table portion of the WCT to indicate the zones where the steps are located. This can be easily accomplished with a preprinted paper form and a pencil. The graph portion of the WCT then treats each zone as a single location. If we compare the hand-drawn sketch in the upper part of Figure B.4 and the sketch in the graph portion of the WCT, we can see that the upper sketch is much simpler. However, it lacks the details that are in the table portion of the WCT preprinted form. Whether you try to modify a preprinted form each time or you draw a table version with a pencil and paper is up to you and your needs. However, if you understand what you are trying to accomplish, you should find that the preprinted form is only a convenience and that if you are constantly modify-

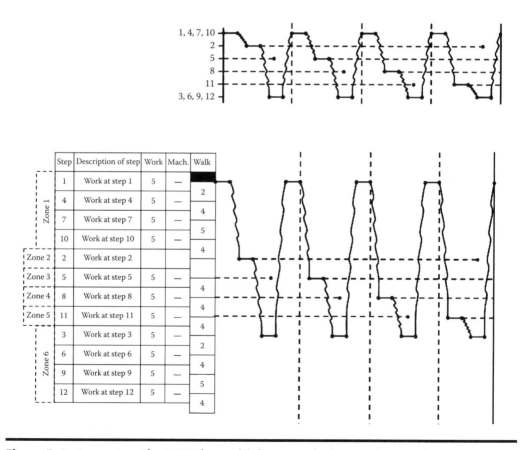

Figure B.4   Impact on the WCT for multiple passes before cycle repeats.

ing them, then you may be better off hand-drawing the entire WCT, table and graph, by pencil and paper.

Summary of the Guidelines

There were a few guidelines developed in the text for assistance with the layout sketch and job design. They are summarized here for convenience:

Basic Layout Sketch:
- Sketch a rough approximation of each work station and its location in relation to the other work stations.
- Each major work sequence step is shown as a number within a circle.
- The steps are connected by solid arrows.
- The return to the step that begins to repeat the sequence is shown with a dashed arrow.
- Work step, equipment name, etc., can be shown for clarity.
- Parallel tasks that do not have to be performed in a particular sequence compared to other steps but simply need to be performed on demand are denoted by letters.
- A combination of sequential and parallel tasks is denoted by using numbers, letters, or a combination of both.

Job Design:
- Minimize walk between adjacent worker locations.
  - Watch for angles <90° when changing direction and allow time for abrupt changes in direction (see Figure 5.18).
  - Make stations as narrow as possible (you can always add space if needed later).
  - Eliminate or minimize gaps between process locations.
  - Arrange the processes in a pattern that allows transfer of work but supports smooth product flow.
- Avoid walking back over ground just covered when not transporting a part.
- Minimize empty-handed walk when the worker is the means of transport in the cell.
- If possible, the same worker should handle the first and last processes to link the product going into the cell with the product leaving the cell. (This will give the cell a more consistent pacing mechanism to meet the takt or desired cycle time.)

- ▪ Use diagrams to represent the walk planned for each worker.
  - – Dashed lines indicate return to starting point—expected empty-handed walk.
  - – Heavy lines indicate planned walk to a nonadjacent process location.
  - – Normal lines indicate planned walk to an adjacent process location.
  - – Assume that there are fixed handoff points between adjacent workers.

Examples of Work-Combination Table Issues

The following figures in this appendix illustrate some common issues and show how the WCT can be used to reveal them. Most problems can be discovered using a combination of the techniques illustrated in Figures B.5–B.14.

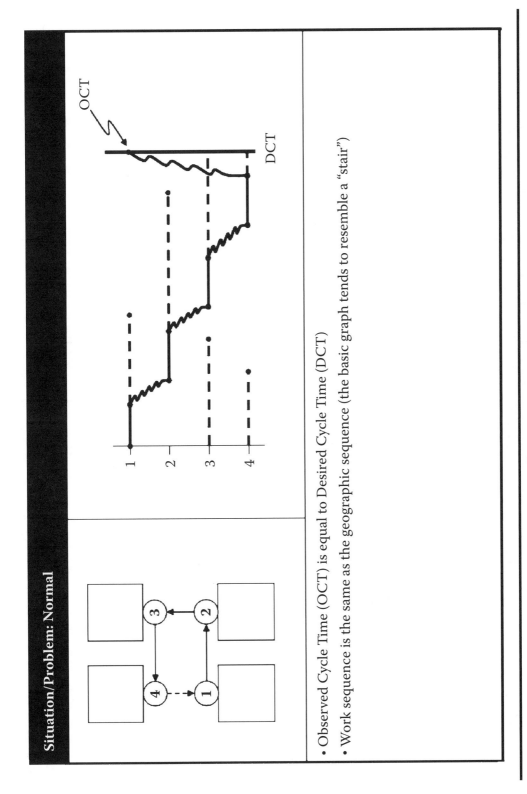

Figure B.5   Layout sketch and WCT graphic: normal.

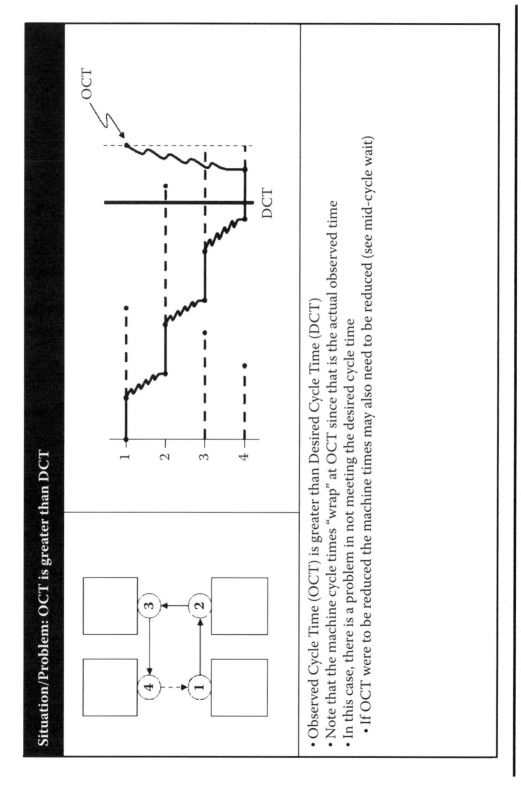

Figure B.6 Problem or issue: OCT is greater than DCT.

Situation/Problem: OCT is greater than DCT

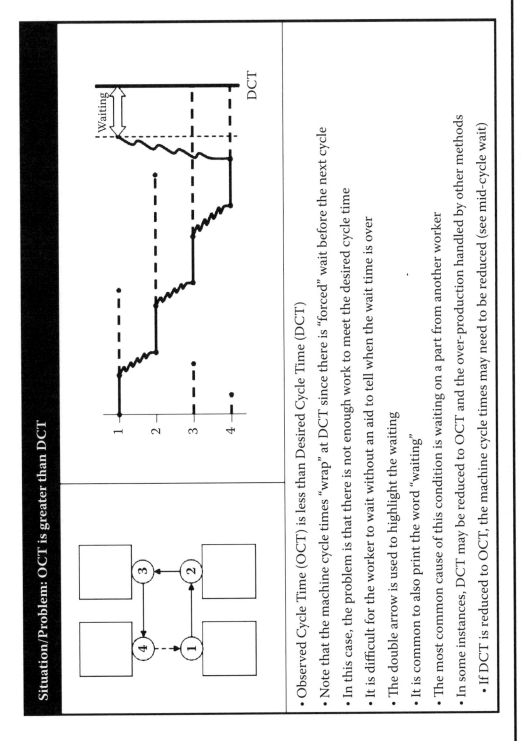

- Observed Cycle Time (OCT) is less than Desired Cycle Time (DCT)
- Note that the machine cycle times "wrap" at DCT since there is "forced" wait before the next cycle
- In this case, the problem is that there is not enough work to meet the desired cycle time
- It is difficult for the worker to wait without an aid to tell when the wait time is over
- The double arrow is used to highlight the waiting
- It is common to also print the word "waiting"
- The most common cause of this condition is waiting on a part from another worker
- In some instances, DCT may be reduced to OCT and the over-production handled by other methods
- If DCT is reduced to OCT, the machine cycle times may need to be reduced (see mid-cycle wait)

Figure B.7   Problem or issue: OCT is less than DCT.

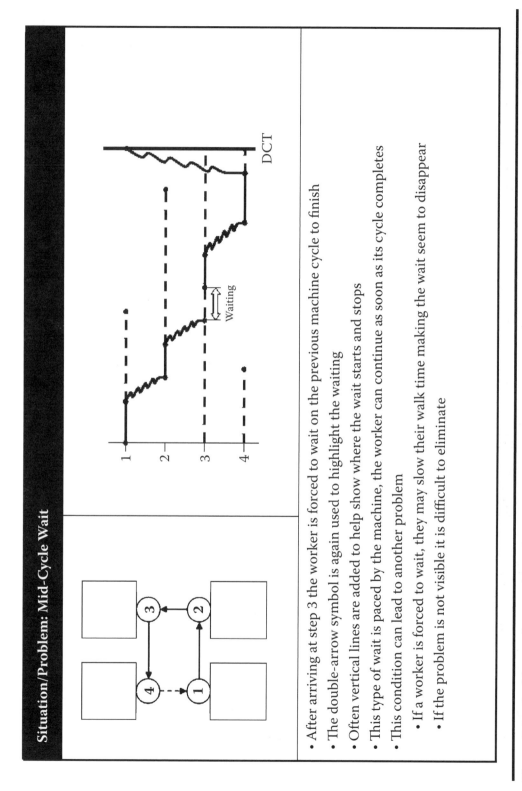

Situation/Problem: Mid-Cycle Wait

- After arriving at step 3 the worker is forced to wait on the previous machine cycle to finish
- The double-arrow symbol is again used to highlight the waiting
- Often vertical lines are added to help show where the wait starts and stops
- This type of wait is paced by the machine, the worker can continue as soon as its cycle completes
- This condition can lead to another problem
  - If a worker is forced to wait, they may slow their walk time making the wait seem to disappear
  - If the problem is not visible it is difficult to eliminate

Figure B.8   Problem or issue: mid-cycle wait.

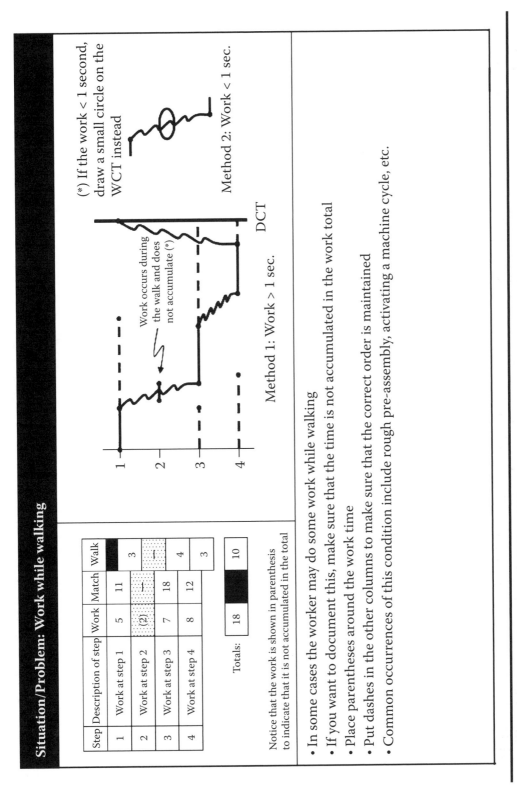

Situation/Problem: Work while walking

Step	Description of step	Work	Match	Walk
1	Work at step 1	5	11	3
2	Work at step 2	(2)		
3	Work at step 3	7	18	4
4	Work at step 4	8	12	3

Totals: 18 | 10

Notice that the work is shown in parenthesis to indicate that it is not accumulated in the total

Work occurs during the walk and does not accumulate (*)

Method 1: Work > 1 sec.

DCT

(*) If the work < 1 second, draw a small circle on the WCT instead

Method 2: Work < 1 sec.

- In some cases the worker may do some work while walking
- If you want to document this, make sure that the time is not accumulated in the work total
- Place parentheses around the work time
- Put dashes in the other columns to make sure that the correct order is maintained
- Common occurrences of this condition include rough pre-assembly, activating a machine cycle, etc.

Figure B.9   Problem or issue: work while walking.

Situation/Problem: Walk is equal to zero

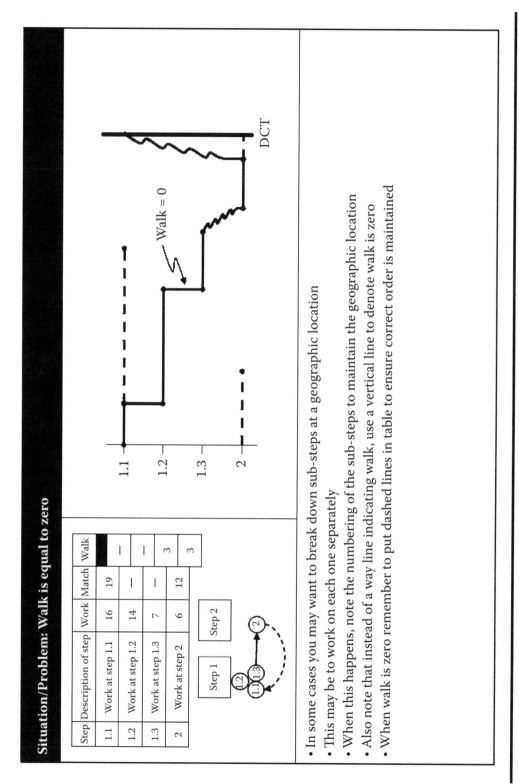

Step	Description of step	Work	Match	Walk
1.1	Work at step 1.1	16	19	
1.2	Work at step 1.2	14	–	–
1.3	Work at step 1.3	7	–	–
2	Work at step 2	6	12	3

- In some cases you may want to break down sub-steps at a geographic location
- This may be to work on each one separately
- When this happens, note the numbering of the sub-steps to maintain the geographic location
- Also note that instead of a way line indicating walk, use a vertical line to denote walk is zero
- When walk is zero remember to put dashed lines in table to ensure correct order is maintained

Figure B.10   Problem or issue: walk is equal to zero.

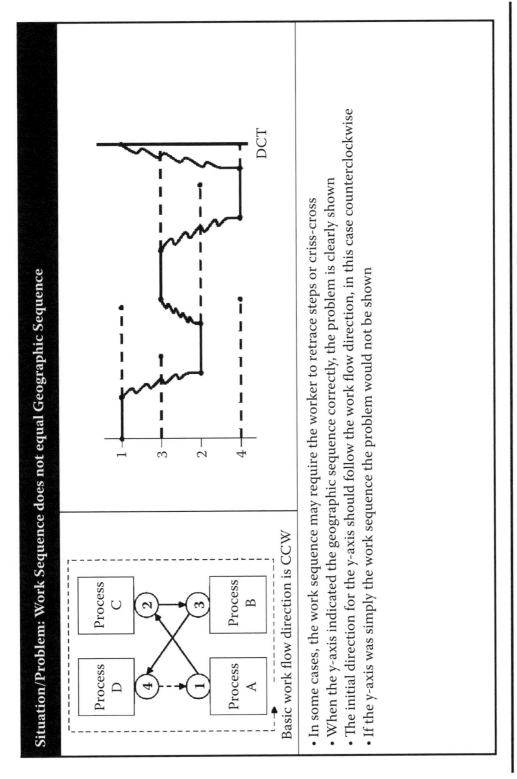

Figure B.11 Problem or issue: work sequence does not equal geographic sequence.

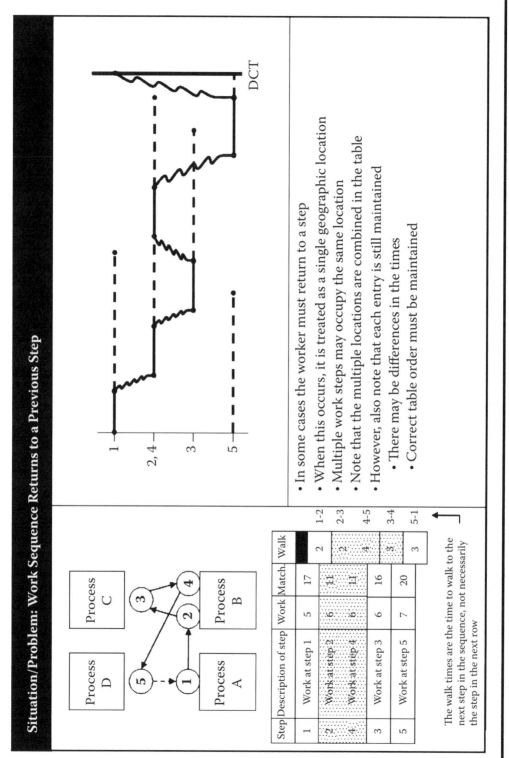

Figure B.12   Problem or issue: work sequence returns to a previous step.

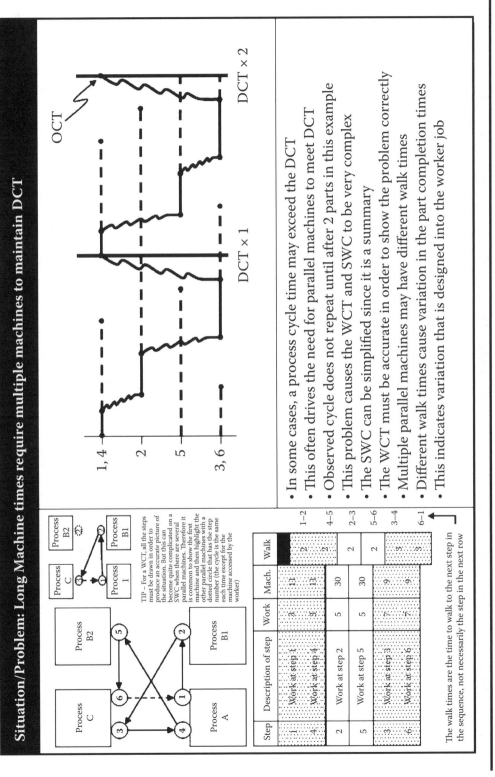

Figure B.13   Problem or issue: long machine times require multiple machines to maintain DCT.

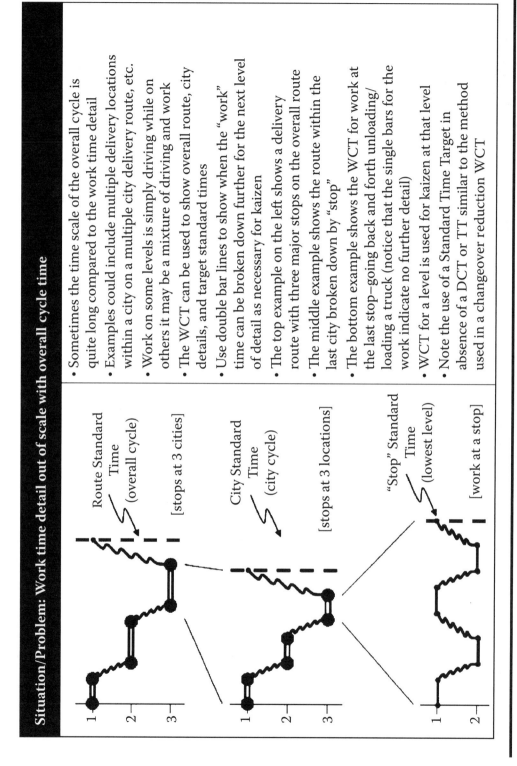

Situation/Problem: Work time detail out of scale with overall cycle time

- Sometimes the time scale of the overall cycle is quite long compared to the work time detail
- Examples could include multiple delivery locations within a city on a multiple city delivery route, etc.
- Work on some levels is simply driving while on others it may be a mixture of driving and work
- The WCT can be used to show overall route, city details, and target standard times
- Use double bar lines to show when the "work" time can be broken down further for the next level of detail as necessary for kaizen
- The top example on the left shows a delivery route with three major stops on the overall route
- The middle example shows the route within the last city broken down by "stop"
- The bottom example shows the WCT for work at the last stop–going back and forth unloading/loading a truck (notice that the single bars for the work indicate no further detail)
- WCT for a level is used for kaizen at that level
- Note the use of a Standard Time Target in absence of a DCT or TT similar to the method used in a changeover reduction WCT

Figure B.14   Problem or issue: work time detail out of scale with overall cycle time.

Appendix C: Taking Measurements with a Stopwatch

Using a stopwatch to measure the time it takes for a worker to complete a single work cycle of a repeating task can be very simple. We start the measurement at the point we believe the cycle begins and stop the measurement at the point we believe the cycle ends. When the work cycle continuously repeats, the beginning and ending points should coincide with one another. In other words, the ending read point would be the next occurrence of the starting read point. This is the point where the current cycle ends and the next cycle begins.

If we want to take multiple cycle measurements, the process gets a little more complicated. For example, if we wanted to take 10 worker cycle readings, if we are using a simple stopwatch with no memory or lap functions, the easiest way to take 10 cycle measurements would be to take 10 separate readings. However, in order to do this, it is usually necessary to capture the cycle time by stopping the stopwatch at the end of a cycle and take the reading. If the worker is continuously repeating the cycle, the worker does not stop while this is happening, so getting 10 consecutive cycle measurements in this manner may not be possible, especially as the cycle times get shorter. It might be possible to have multiple people using multiple stopwatches and trying to coordinate who was taking what measurement, but the difference (or overlap) in the time between the starting of one person's stopwatch and the stopping of the other could create too much variation. We could measure the time for a specific number of cycles and then divide the total time by the number of cycles completed in that time, but that only gives us an average time. It does not let us measure the variation between cycles.

However, many stopwatches these days are available with a "lap" function. These were originally intended for activities such as sporting events where a repeated course is used. An example of this would be a person running laps around a track. The stopwatch would be started at the sound of the starting gun and if the event had multiple laps, the lap button would be activated as the runner crossed the starting line to take a time reading for that particular lap while the stopwatch continued to accumulate time for the next lap. This concept works very well for events where there is plenty of time to record the lap time, reset the lap function, and prepare for the next lap all while the time continues to accumulate. But as the cycles get shorter and shorter, it becomes more difficult to take these multiple readings. Today, electronic stopwatches are available with multiple memories so that it is only necessary to hit the start button and then continuously hit the lap button until the last cycle reading, at which time the stop button is hit to end the taking of readings. The measurements for each lap can be retrieved from memory after the readings are taken. This allows the person using the stopwatch to focus on the event more intently.

This type of stopwatch is very easy and convenient to use. It also allows taking multiple cycle measurements where the cycle time is very short. The practical limit is probably the reaction time of the person using the stopwatch. But measuring the total of each cycle is not the only information of interest. Therefore, it is not just the individual times themselves that are of interest for our purposes but, rather, the fact that the lap function stores the accumulated time from the last event where the function was used and can be retrieved sequentially after the fact. This allows us to take individual measurements of the *components* making up the work cycle by simply having easily recognizable starting and ending read points. The "laps" are then the consecutive time components that make up each individual work cycle.

Take for example a work cycle that has four different geographic locations where work is performed and the worker must walk a short distance as he or she moves between the locations in the performance of the tasks. If we quickly sketch out a work-combination table (WCT) with a pencil and paper, we would see that the four work times are separated by the walk times to move to each succeeding location. Because the last time component is comprised of the walk back to the beginning point in this example, there would be a total of eight time components: four work components and four walk components. We also know that it is necessary for the worker to meet the standard time for a cycle each time it occurs in order to stay on track with the desired output. It would be nice to take multiple, consecutive

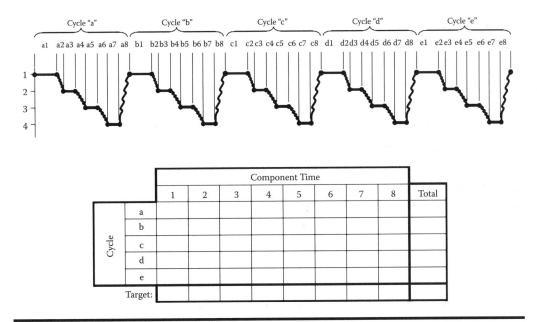

Figure C.1   Capturing the work element data using a stopwatch with multiple lap memories.

measurements in order to get a better representation of the actual work over time. This would also be useful for measuring the variability between cycles. Using an electronic stopwatch with multiple memories would allow the user to gather multiple time components of individual work cycles within the limits of the number of lap memories available. In Figure C.1, there are five complete cycles shown in the upper portion of the illustration. If we had easily recognizable starting and ending read points defined and an electronic stopwatch with multiple lap memories, the five cycles and all the internal time components could be measured and retrieved after the last cycle measurement was taken.

In Figure C.1, we see that there are eight components and five total cycles, so there would be a minimum of 40 memories required for the stopwatch. The process for taking the measurements would be fairly simple and straightforward. The steps might be as follows:

1. Develop a layout sketch.
2. Develop a rough WCT graph/sketch.
3. Identify the time components of the work elements and tasks involved.
4. Determine easily recognizable starting and ending read points for each time component.

5. Develop a simple table to record the time component measurements in the proper order.
6. Reset the stopwatch.
7. Press the Start/Stop button at the beginning of the first cycle.
8. Press the Lap button at each succeeding starting read point.
9. Press the Start/Stop button for the end read point of the last cycle.
10. Retrieve the time measurements from the stopwatch and record them in the proper location in the table.
11. Fill in the total for each cycle by adding the times for the row.
12. Optionally choose a target time for each component utilizing the method you deem most appropriate (lowest repeatable, averaging, hi/lo averaging, etc.).

Once the measurements are retrieved and organized in the correct order, the information can now be used for setting targets, determining variability, and other useful purposes. There are, of course, many purposes for this information, but we felt it was important to introduce the potential of the electronic stopwatch with multiple memories. They are fairly inexpensive, but as the number of available lap memories increases, so does the cost. Therefore, it is important that the user plan out the most effective use of the stopwatch capabilities and use the standardized work tools properly.

Returning to Figure C.1, we notice that there is a table in the lower portion of the illustration. This is just an example of a table developed in step 5 above in order to accommodate the 40 time-component measurements. Notice that in this case the components are recorded in the columns in order for the rows to represent the individual cycles. There is an additional column added in order for the sum for each row to be calculated and the total determined for each observed cycle time (OCT). Similarly, an extra row has been added in order to allow a representative component time to be determined using whatever method is deemed appropriate. It does not matter what the table actually looks like as long as it serves its purpose correctly. It is most important to plan things out so that the measurements taken are retrieved in the correct order so that the information is reflected accurately. It is easy to make a mistake if the proper planning is not performed.

One last point to make is the importance of defining easily recognizable starting and ending read points. If the points are not very definite, the opportunity for variation can be increased due to reaction time, human error by the user, and other similar factors. Sounds, motions, lights, and other definite events that can serve as an absolute point prove very useful for this purpose. But there may be times when it is difficult to establish an easily recognizable read point. In such cases, it may be better to just combine a couple of time components so that the points used are very definite and distinct. This is why it is important to know how to properly sketch a layout and work-combination table with a pencil and paper so that the data taken matches the sketches. If time components must be combined to get definite starting and ending points, this should be reflected on the rough WCT sketch. This way, kaizen can continue rather than wasting time trying to obtain data that may not have as great an effect on the outcome as the resources expended to obtain it. One final comment on taking measurements in developing standardized work: Practicing at the work site with the stopwatch before you start will make you more confident and should be able to reduce human error by the user.

Appendix D: Workplace Mock-Ups and Simulation Philosophy

Several times, we have mentioned the importance of keeping the improvement changes small. The main reason is so that each change can be evaluated in its own right without the chance for unintended effects of other concurrent changes affecting the outcome. If the change is kept very small, and there are no other simultaneous changes made, the results are much easier to verify. However, it is just as important that the *expected* results be just as simple and clear. It is not enough to start making changes simply because it seems like a good idea or that it "couldn't hurt." This is not unlike an improvement workshop where a list of "action items" are generated but not implemented at the time. If the changes are not well thought out and verified against the expected results, then this action item list is simply a list of things that *might* be good ideas.

Although the main reason to keep the change small is so that it can be evaluated in its own right, there are other reasons that are also important. Another reason that comes to mind is so that the change can be tried and verified quickly and with very little resources expended. If the results are deemed good, then the next change can be tried quickly afterwards and the process continues. However, with the emphasis on eliminating waste, it is important that we not risk too much time and resources on trying things that do not provide the expected results. But it is also important to note that if we are not having some improvement ideas that fail to meet or exceed the expected results, then we are probably not taking enough risk. It is also a good idea to not have to spend a lot of time analyzing the results. We do not want the so-called analysis paralysis to develop. By now it should be quite obvious that the smaller the change, the easier it is to verify the results against the expectations. But some changes do require some analysis, so it is best to keep them as small as possible. Remember that the kaizen attitude should support continuous improvement, and when the changes are small, the disruptions and variation that will occur temporarily will also be kept small. There are many other reasons that could be given to support the case for keeping the changes small. However, it is not the purpose of this section to justify the merits of small changes but rather to discuss the philosophy of developing and verifying changes, especially when the change involves methods and motions.

Sometimes it is not practical to develop ideas right at the workplace. This could be due to limited space, production schedules, cleanliness issues, safety, etc. But this should not stop us from working on improvements. One of the ways that can help to develop improvements quickly and inexpensively is by using a workplace mock-up. A workplace mock-up is a three-dimensional (3-D) representation of the workplace where the improvement is desired. Similar to the rough sketching techniques used for developing layout sketches and work-combination tables with pencil and paper, this 3-D mock-up does not have to be to perfect scale, although it should be a reasonable facsimile. Also, the mock-up does not have to be made using expensive materials and processes. Often a reasonable mock-up can be built using discarded cardboard boxes, packing materials, shipping tubes, and other "junk" found around the office or company. Using common supplies such as sturdy tape, safe scissors, safe box cutters, and tape measures, the discarded materials can be very quickly manipulated into a rough 3-D representation of the workplace, workstation, or machine where the improvement is desired. Again, similar to the rough sketching techniques, it is not necessary to get the 3-D mock-up to perfect scale, but it is a good idea to get it close so that it suffices for our purposes of analyzing motions and methods changes.

Also, when using workplace mock-ups, we do not need to put all the details into the "model" we are building, only the details that are required. Often this is the width, depth, and correct work-surface height and other details that are relevant to the improvement involved. For example, if we wanted to build a mock-up of a drill press in a workshop, the details that we might find important include the width, depth, and height of the work table, a representation of the work holder, the arm that moves the drill up and down, the controls, etc. It is not always necessary to make a "working" model, although we have seen some very accurate representations put together very quickly with nothing but cardboard, tape, and materials found in a dumpster. Because the materials are inexpensive and easy to work with, you are limited only by your imagination. Remember that the purpose is to be able to simulate the improvements, so not all the details are necessary, but make sure that the workplace mock-up is sturdy enough to survive a little usage and movement.

When we consider how sturdy the workplace mock-up should be, consider how it will be used. If you are going to use it to develop standardized work, it may only be necessary to focus on the details that affect the time components of the job (work, walk, and wait). The work surface height, workstation width and depth, or other "footprint" type of dimension that might impact the worker's motions seems to fit this category. These motions might be due to reaching for materials being applied or used at the workstation, any tools that the worker would need, performance of the work itself, activation of any controls, etc. However, consider that in order to use the workplace mock-up properly, it may be necessary to move some things around, so not only does the workplace mock-up need to be fairly sturdy to survive some moving and rearranging, it is important that things be easy to modify as well. Therefore, there is a balance between sturdiness and flexibility that is needed to build a useful workplace mock-up quickly.

It is often necessary to build workplace mock-ups of more than just a single workstation, especially when the walk time component is being considered. For example, if we were going to build a workplace mock-up for the toy truck example that we discussed in Chapter 3, we would want to consider all four workstations. This would require us to know some of the specifics about the dimensions and locations of the workstations if they already existed and were in use. Although the information might be available from a scaled layout, if would be best to go directly to the place where the work happens and get the rough measurements yourself, if possible. The reason for this is simple: Layouts can become outdated, but being right at the workplace will allow you to gather information that will not be available on the layout. Some of this information could be critical to your improvement efforts. For example, you may observe containers, tools, materials, scrap, or personal property such as radios, jackets, coffee mugs, etc. Certainly not all of these would be shown on a layout, but it is obvious that that could definitely have an effect on the work. Also note that it may be helpful to move the workstations around, making them closer together or in some cases a little farther apart. This is another reason for considering sturdiness.

Sometimes a workplace mock-up may be needed over a longer period of time. This could be because of limited work space that must be shared with other improvement efforts, or because the team can only work on the improvements for short periods over a longer period of time, or for other similar reasons. But so far we have been discussing modeling workstations that already exist. Workplace mock-ups can be a very powerful tool to use for not only developing standardized work, but also for developing the equipment and workstations themselves. Consider that when the equipment and workstation already has been purchased or constructed, our opportunities for improvement become more limited because the equipment and workstation *drive* the standardized work design. When they have not yet been procured, we are presented with the opportunity to try and work on developing a preferred method that can then *drive* the design or selection of the equipment and workstation. This is one of the most powerful aspects of the workplace mock-up, because it allows the kaizen attitude to be applied even earlier in the design stages of a project. And since this can happen quite early in a project, the workplace mock-ups may need to be around for quite some time. If they are not at least somewhat sturdy, they may not survive the design stage of the project.

Another important point about workplace mock-ups is their value in communicating ideas. Often it may be difficult to put into words or sketch simple enough drawings to make someone's ideas clear to everyone. However, when dealing with workplace mock-ups, this can often be simply a matter of modifying the mock-up or building an alternative version. Because this can be done quickly and inexpensively, many ideas can be tried that might not otherwise get sufficient consideration. Sometimes an idea might not work as stated, but if the team gets the opportunity to see the idea, there might be other ideas built on the original idea, and this can be a very powerful tool for supporting the development of offshoot ideas. It also gives the team member who originated the idea the assurance that his or her idea was given sufficient consideration and can therefore also be a great team building tool.

We have seen many workplace mock-up projects where managers and even executives rolled up their shirt sleeves and got involved in the design and building of the mock-ups. Although a workplace mock-up seems like a simple activity, it can have a profound positive impact on the team and their efforts. However, it is also important to remember that the primary purpose of the activity is not to produce a workplace mock-up. It is to use the workplace mock-up to develop ideas for improvement. This improvement can be for the standardized work for an existing job or the design of good standardized work for a job that has not yet come into existence. The improvement can also be for the equipment or the workstation that already is being used in a job, or it can be used to design and develop the equipment and workstation for a job that is being designed so that they support the pre-ferred standardized work. But the workplace is not the only extent of the uses for the mock-up. Mock-ups can be used for product designs, tooling designs, fixture or work holder development, and in many other ways. It is limited only by your imagination. Often a mock-up activity is part of a workshop or formal improvement event. These types of activities have tre-mendous team-building potential as well as the benefits we have discussed so far. When there are a lot of people available, it is entirely feasible to build a mock-up of an entire system, whether manufacturing, transactional, or other type. We highly recommend using such opportunities to get the teams engaged and involved in the design of new systems as well as the kaizen of existing systems by using the workshop concept.

A mock-up provides the opportunity to quickly get an idea of the rami-fications or impact that a change may have on the methods and motions. It has the flexibility of the pencil and paper sketch with the added benefit of the three-dimensional view. This can be very useful in dealing with physi-cal motions such as hands and feet. But there are other motions or aspects that may be of interest to us in our efforts to improve. For example, the movements of the worker's eyes, multiple reaches or motions, grasping and orientation of materials or hand tools, etc. Some of these, such as the pre-sentation of materials and hand tools, can simply be modeled themselves using the same techniques as we used to develop the rough workstation or equipment. This is why a mock-up activity can quickly turn into a cardboard model of the workplace. This often proves very useful, because not only does this give the participants the chance to get a 3-D perspective of the situation, but it also allows people who might actually be doing the work to have positive input to the development of the improvement, whether it is for the standardized work, the equipment, or the workplace. This will also give you a very good perspective to evaluate any safety and ergonomics concerns that may be avoided by design.

A technique that can be used to help understand the worker's eye move-ments, at a mock-up or at the actual workplace, involves thread, tape, and scissors. It is best to use a thin thread such as used for sewing and the transparent tape found in the office. The idea is to go through the job requirements and establish where the worker's eyes must move during the course of the cycle. For example, if a worker must look up to a control panel to see if the light indicates that the part has tested good, then must reach to the right to grasp a label from a printer, then apply the label at a specific point and orientation on the product, the worker's eyes will have to move back and forth. If we were trying to measure or capture these movements in some way in order to study or improve them, this method can be used to visually represent the movements. The idea is simple: Using the tape, attach the start of the thread at the beginning location, then string the thread to the next location and attach it with tape, and continue stringing the thread and affixing with tape at each point that the worker's eyes must move to until the cycle is complete. This can often look like a spider web, but it can be very effective. It works well for mapping eye movements, but it can also be used to show the path that the worker's hands follow during the motions they make to perform the job. This is especially helpful when there are a lot of hand movements through the same general area to assemble, grasp or replace hand tools, grasp components, etc.

Appendix E: A Little More on the Kaizen Attitude

We have mentioned the kaizen attitude quite a bit throughout this book. The purpose of this appendix is to discuss this concept in a little more depth. The kaizen attitude is not just making small changes instead of large changes; it is more of a philosophy or culture where the main purpose is to always seek to improve while balancing this goal with stability and minimum variation. Whenever change is involved, there will be uncertainty. This can cause many people to have a difficult time adapting to the idea of change, and therefore they can often be negative about the concept, perhaps without even realizing it. As a reminder, the kaizen attitude must consider and maintain a safe environment for the worker to operate in. At the same time, the quality of the product or service delivered must not be compromised.

One of the most prevalent examples of the uncertainty in trying to support the kaizen attitude that we have observed often starts from the standpoint of someone explaining all the reasons why something cannot work when confronted with a new idea or suggestion. There is nothing wrong with listing the problems with a new idea or suggestion, but if the person takes the defensive stance against change, it can quickly spread throughout a group of people and slow or even stall the improvement process. This can basically be summarized as a "this won't work because..." approach. However, in the kaizen attitude, even if the idea as suggested has many obstacles or problems, there still may be some improvements that can be derived from an offshoot idea, or at least from looking at things from a different perspective than before. This different perspective might be summarized as the "what would need to change to make this idea viable?" approach. Both approaches deal with the problems and obstacles, but the latter approach has a positive problem-solving attitude that often tends to get a group of people all looking for solutions rather than looking for weaknesses in another person's arguments why something might not work, which is often observed in the first approach.

It has been our experience that a positive attitude for change can often lead to breakthrough improvements over time. We have facilitated or participated in hundreds of workshops and improvement events around the world over the years and have noticed that when the negative approach is prevalent, it often tends to waste the limited and potentially expensive resources of the group working on improvements. Usually we observe some of the participants trying to explain why an idea is not feasible or cannot work, while others are trying to explain why it could work if the conditions were different, that there are flaws in another person's argument, and so forth. This not only slows things down, it can actually cause friction among the people trying to work together. This can often revert to simple arguing rather than constructive discussion of problems and benefits. When the positive approach or attitude is prevalent, it usually gets most, if not all, of the group moving in the same direction. It is critical that the people involved with the improvement process embrace the positive approach to support the kaizen attitude, both workers and management.

If we stop and consider for a moment the problems that can arise from the negative approach, we can see that there are a lot of social aspects to the process. For example, in the heat of debate, it can sometimes become easy for tempers to flair. The intent of the kaizen attitude is not to make people's lives more difficult, but rather to improve things. This leads us to another issue that we have found that is very important to the kaizen attitude, which is respect for the people involved. This does not refer only to the workers being affected but also to the company. It is a commonly held opinion that most companies only think about the bottom line and not how improvement efforts impact the people who work for them. Although our own observations have seen evidence to support this opinion, this is not always the case. Often the real reasons behind the hard decisions being made are not communicated to those whose livelihoods are being affected. Communicating the reasons behind hard decisions may not change the end result, but it can definitely have a better effect upon those involved. For example, consider a company on the verge of bankruptcy, where it is essential to reduce expenses and costs immediately to remain in business. If this fact is kept from the people who are asked to make changes or come up with cost-saving ideas, the sense of urgency is being lost.

Due to the commonly held opinion that businesses are looking only at profits and not at their employees, an improvement initiative can often be misinterpreted as simply a grab at more profit. On the other hand, it is the company's right to try to reduce waste at every opportunity. In an effort to keep the real situation hidden, many companies will simply inform the teams involved that certain savings are needed, but no specifics are shared. We have seen this culminate in a great kaizen event where the team, who were extremely excited and enthusiastic about their results, is confused by the unimpressed management feedback. Often, this is because management did not share the true reasons for the workshop or event, but rather left vague savings goals. Sometimes it seems necessary to keep this information from the team, especially if the use of certain financial information may create confusion, but we have observed that it is best to inform the team of the full seriousness of the situation. If the intent is to reduce the department with 20 workers by two workers, it is much more palatable to participate when your attitude is that you are trying to save the jobs of those 18 who are left.

There are basically two paths that companies can take when faced with these type of decisions in their efforts to reduce waste. Suppose we have a situation where the savings event resulted in the reduction of labor required by 50%. One path is to simply reduce labor by 50% and take the savings. The company is now able to produce the same amount of work output but with half the labor costs. It is the right of the company to do this if they so choose. The other path is to look at the savings from another perspective. The company can now produce twice the work output using the same 20 people as they could before.

The first path is a very tempting way to boost profits in the short term. The fact that this does happen in some industries can reinforce the opinion that businesses are only interested in profits and therefore are inherently greedy. It also has a negative impact on the workers involved, not only the two who might be laid off, but on all those who participated in the improvement effort. And how enthusiastic will the participants in the next improvement initiative be? Even if the reduction is necessary to keep the business open and ensure the jobs of the remaining 18 people, it is extremely difficult if all the facts are not shared with the workers. It is much easier to have a positive attitude when there is an environment of trust, even when the situation is serious. So in our observations, the real issue is trust— not sharing all the facts. This is similar to a lesson we learn from farmers: Do not eat your seed corn in the winter or you will not have corn to plant in the spring. Although it saves money, it does not provide for the future. If the people are given all the facts, it is much easier to live with the results of the savings efforts in business-survival situations. After all, most people in the workforce, regardless of the level of education, have capable levels of awareness and common sense. The key is that if the kaizen attitude is truly prevalent throughout the company, the only time radical measures would be necessary are under catastrophic conditions.

The second path is not always practical, especially when we consider the concept of value. For example, if the workers in the department were producing products that can be sold, this opens up the opportunity of trying to increase business to utilize the newly found capacity. The sales department could go and try to get new business by offering lower prices. Even if the market could not support new customers, it might be possible to win some business back from the company's competitors. And the displaced workers could be trained to help spread the kaizen attitude in other parts of the company, thus potentially multiplying the benefits. However, what if the department with the 50% reduction in labor costs did not produce products but instead were administrative in nature? It is easy to revert back to the first-path thinking if we believe that the department did not produce value for the customer. It is important to remember that the concept of value is not as simple as black and white; the incidental work infuses a gray area because it represents work that may or may not add value, but is still necessary at the moment. In situations such as the administrative department example, it may be incidental work that is being impacted, and the decision about value or nonvalue will have to be made. But that is for the work; the people are still involved, and the company must decide which course to pursue. The same options are available to the company as before. The extra capability could be used to bring in other administrative tasks that are currently being performed by outside companies. It could also be used to do something else entirely. The key is that this should be looked at as an opportunity to grow rather than shrink.

Before we conclude our discussion on our point about respect for the people involved, we want to note that this is not a one-sided point. The people working for the company must also respect the management and the owners. It is still *their* company, and the workers are working for them. Companies exist for a purpose, and although there are many purposes why a company might exist, one of the most prevalent is to produce a profit. Our observations also support that there is a growing opinion out there among business owners and upper management that the workers are only concerned with getting more pay and benefits. Again, this is not always the case, but it is similar to the opinions about businesses and companies discussed earlier. The workers must understand that in order for the company to survive and provide benefits and wages, they must make a profit. If they are to take the considerable risk to invest and try to grow, providing more jobs and possibly better pay and benefits, the rewards must be worth that risk. This is what we meant when we mentioned the symbiotic relationship between the workers and the company. There must be rewards for both, but conversely that means there will be risks for both as well. The key is to make both parties a part of the kaizen process. Therefore, it is important to remember that the company must be profitable while the workers are secure. If either party seeks to increase their side of the equation to excess, it will adversely affect the other side. If they are increased together, slowly but continuously, not only do both sides win, but often the business-survival situations that can affect many companies can often be avoided entirely.

References

Dennis, Pascal. 2002. *Lean production simplified: A plain-language guide to the world's most powerful production system*. New York: Productivity Press.

Goldratt, Eliyahu. 1990. *The Haystack Syndrome: Sifting information out of the data ocean*. New York: North River Press.

Kitano, Mikio. 1997. *Toyota production system: One-by-one confirmation*. Keynote address, University of Kentucky Lean Manufacturing Conference, May 15, 1997.

Ohno, Taiichi. 1988. *Toyota production system: Beyond large-scale production*. New York: Productivity Press.

Shingo, Shigeo. 1989. *A study of the Toyota production system: From an industrial engineering viewpoint*. Translated by Andrew P. Dillon. New York: Productivity Press.

Index